中国建筑学会室内设计分会
北京建筑大学 编

中国建筑学会室内设计分会推荐专业教学参考书

Green
Building Indoor Assembly Design

绿·造卷——建筑室内装配化设计

中国建筑学会室内设计分会推荐专业教学参考书

『室内设计 6 +』2019（第七届）联合毕业设计

中国水利水电出版社
www.waterpub.com.cn
· 北京 ·

图书在版编目（CIP）数据

绿色营造卷：建筑室内装配化设计 / 中国建筑学会室内设计分会，北京建筑大学编. -- 北京 ：中国水利水电出版社，2020.1
中国建筑学会室内设计分会推荐专业教学参考书 "室内设计6+" 2019（第七届）联合毕业设计
ISBN 978-7-5170-8135-7

Ⅰ．①绿… Ⅱ．①中… ②北… Ⅲ．①住宅－室内装饰设计－毕业设计－高等学校－教学参考资料 Ⅳ.
①TU241

中国版本图书馆CIP数据核字(2019)第246415号

书　　名	中国建筑学会室内设计分会推荐专业教学参考书 "室内设计 6+" 2019（第七届）联合毕业设计 绿色营造卷——建筑室内装配化设计 LÜSE YINGZAO JUAN—JIANZHU SHINEI ZHUANGPEIHUA SHEJI
作　　者	中国建筑学会室内设计分会 北京建筑大学 编
出版发行	中国水利水电出版社 （北京市海淀区玉渊潭南路 1 号 D 座　100038） 网址：www.waterpub.com.cn E-mail：sales@waterpub.com.cn 电话：（010）68367658（营销中心）
经　　售	北京科水图书销售中心（零售） 电话：（010）88383994、63202643、68545874 全国各地新华书店和相关出版物销售网点
排　　版	中国建筑学会室内设计分会
印　　刷	天津嘉恒印务有限公司
规　　格	210mm×285mm　16 开本　16.25 印张　855 千字
版　　次	2020 年 1 月第 1 版　2020 年 1 月第 1 次印刷
印　　数	0001—2000 册
定　　价	120.00 元

编委会
Editorial Committee

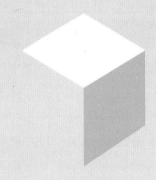

北京建筑大学 BUCEA ACADEMY OF DESIGN AND ART 设计艺术研究院

出版支持：北京建筑大学一流专业建设（市级）项目

内容提要

装配式建筑是用预制部件在工地装配而成的建筑。发展装配式建筑是建造方式的重大变革，是我国当前推进供给侧结构性改革和新型城镇化发展的重要举措，有利于节约资源能源、减少施工污染、提升劳动生产效率和质量安全水平，有利于促进建筑业与信息化工业化深度融合、培育新产业新动能、推动化解过剩产能。

进入新时代，我国更加重视建筑工业化体系建设，坚持标准化设计、工厂化生产、装配化施工、一体化装修、信息化管理、智能化应用，将发展装配式建筑作为重大发展战略，提高技术水平和工程质量，促进建筑产业转型升级。2016 年，国务院办公厅发布《关于大力发展装配式建筑的指导意见》，提出了要「牢固树立和贯彻落实创新、协调、绿色、开放、共享的发展理念，按照适用、经济、安全、绿色、美观的要求，推动建造方式创新，大力发展装配式混凝土建筑和钢结构建筑，在具备条件的地方倡导发展现代木结构建筑，不断提高装配式建筑在新建建筑中的比例。坚持标准化设计、工厂化生产、装配化施工、一体化装修、信息化管理、智能化应用，提高技术水平和工程质量，促进建筑产业转型升级」的发展指导思想。为此，中国建筑学会室内设计分会将「绿色营造——建筑室内装配化设计」作为 2019（第七届）「室内设计 6+」联合毕业设计特色教育创新项目的总命题。

「室内设计 6+」联合毕业设计是由中国建筑学会室内设计分会于 2013 年创立的教学交流活动，2018 年经中国建筑学会批准为室内设计特色教育创新项目。本届活动由同济大学、华南理工大学、哈尔滨工业大学、西安建筑科技大学、北京建筑大学、南京艺术学院、浙江工业大学作为全国活动的「核心高校组」成员，联合多家知名企业共同指导毕业设计。来自建筑学（室内设计）、环境设计、产品设计、艺术与科技等专业的 2019 届毕业生们，通过学会、高校、企业、专家间的协同，就「绿色营造——建筑室内装配化设计」总命题下的多个子课题，联合开展了综合性实践教学活动，从多专业领域感受《关于大力发展装配式建筑的指导意见》提出的「健全标准规范体系、创新装配式建筑设计、优化部品部件生产、提升装配施工水平、推进建筑全装修、推广绿色建材、推行工程总承包」重点任务，促进了毕业设计教学水平和人才培养质量的提升。据此，中国建筑学会室内设计分会组织项目的参与高校共同编写了室内分会推荐专业教学参考书：「室内设计 6+」2019（第七届）联合毕业设计《绿色营造卷——建筑室内装配化设计》。全书分为项目规章、开题调研、中期检查、答辩评审、教育研究、专题讲坛、风采定格等特色章节，主要内容采用中英文对照方式，记录了本届特色教育创新项目的开展情况，图文并茂，内容翔实。

本书可供建筑学、环境设计、产品设计、艺术与科技，以及城乡规划、风景园林、城市地下空间工程、视觉传达设计、公共艺术等专业人员及设置相关专业的院校师生参考借鉴。

序
Preface

苏丹

中国建筑学会室内设计分会
理事长

President of the China Institute of
Interior Design

清华大学美术学院 教授

A Professor at School of Fine Arts,
Tsinghua University

出　师　表

Memorial to the Throne for an Expedition

　　看到了持续六年的"室内设计 6+"联合毕业设计特色教育创新项目以及其累计的丰硕成果，我突然想到了《出师表》。也许有人会说相隔了 1800 年的时光，又是两种类型的事情，一个是出征前的资政，一个是学术活动的总结。它们之间有什么联系呢？我说：当然有！

　　首先触动我的，是我们这个学术组织发展的历程和初心。1989 年，中国建筑学会室内设计分会的前身中国室内建筑师学会成立，在当时这是一件非常具有远见卓识的大事情，事实也如那些开疆拓土的前辈们预料的一样，室内设计这个学科领域、这个行业、这项事业在过去的 29 年中已是江河万里、沧海桑田。用巨变来描述毫不为过，而更加重要的是质变，即"室内设计"逐渐地完善也是一个基本事实。数以百万计的室内设计师，服务着数以亿计的人民大众；每年数以百亿的设计产值，同时撬动着数以万亿计的国民经济产值；此外，室内设计正在迅速地提高人民的生活水平，不断拉近我们和文明的距离。在学术层面，一方面，它的工程属性得到了重视和发展，这得益于几十年来学会所推动的建筑系统性；另一方面，它的文化和艺术属性不断加强，这也得益于学会倡导的包容性。

　　然而在一个逐利的时代，面对这样的一个潜力无穷又商机无限的领域，不同的群体从各自的出发点假以各式各样的口号发起不同的组织。这些组织如飘浮的冰层，在洋流的作用下，在同一领域经常彼此摩擦、相互撞击。另一方面，同样的一群设计师经常拥有多个身份，在不同的浮冰之间来回穿梭。这种群雄并起逐鹿天下的局面，总令我想起《出师表》开篇的那些对天下格局的概述："今天下三分，益州疲敝，此诚危急存亡之秋也。"

　　近几年这种忧虑越来越重，同质化的竞争在多个方面严重影响着这个领域的健康发展。

　　《出师表》第二段"然侍卫之臣不懈于内，忠志之士忘身与外者，盖追……"，这句话谈到了"坚守者"。这

I thought of *Memorial to the Throne for an Expedition* at the sight of the Innovation Program of Characteristic Education on "Interior Design 6+" Joint Graduation Project and its fruitful achievements. It can be argued that they are 1800 years apart and two types of things, one being a memorial to the throne before going into battle, the other being a summary of an academic activity, so what's the connection between them? I would answer, "Of course there is a connection between them!"

The first thing to touch me is the development history and original aspiration of the academic organization. The China Institute of Interior Design (CIID), formerly known as the Society of Chinese Interior Architects, was founded in 1989 with farsightedness and insight. As expected by our predecessors that broke the ground, as a discipline, an industry and an undertaking, interior design has achieved great development and great results in the past 29 years. Its development can be described as a radical change, and what matters the most is the qualitative change, in other words, it is also a basic fact that "Interior design" has been gradually perfected in essence. Millions of interior designers serve billions of common people; interior design has an annual output value of tens of billions of yuan, and makes an indirect contribution to multi-trillion-yuan national economic output; besides, interior design is rapidly improving the people's living standards and shortening the distance between civilization and us. At the academic level, on the one hand, its engineering attributes have been valued and developed. This benefits from the architectural systematization that has been promoted by the ASC in the past several decades; on the other hand, its cultural and artistic attributes are being strengthened. This owes to the inclusiveness espoused by the ASC.

However, in this money-oriented era, faced with such a field with infinite potential and unlimited business opportunities, different groups have founded different organizations with different slogans from their own perspectives. These organizations, like floating ices, often rub and collide against one another in the same field under the action of ocean currents. On the other hand, a designer may have many identities and shuttle back and forth among different floating ices. The contention among different design schools reminds me of the introductory overview of the world situation in Memorial to the Throne for an Expedition: "Today, the empire remains divided in three, and our very survival is threatened".

In recent years, this has been increasingly worrisome for us, and homogeneity competition has seriously affected the sound development of this field in many ways.

The second paragraph of Memorial to the Throne for an Expedition says,"Yet still the officials at court and the soldiers throughout the realm remain loyal to you, your majesty. Because they remember the late emperor…" This sentence mentions "the defender". It reads moving. You may realize that in this era of chaos, there are still people holding fast to the academic value, doing research independently, and working on sacred

是让人非常感动的地方，你会觉得在这个纷乱的时代依然有一群人坚守学术的价值，去做孤独的研究、做神圣的教育。"室内设计6+"联合毕业设计特色教育创新项目就是这种类型的一种鲜明代表，这个项目的主导者就是时代的坚守者。长期以来，中国建筑学会室内设计分会关注设计教育，而这个领域竟成为一个行业热点中的高海拔地区。高处不胜寒，以此见真心。六年以来，项目核心高校室内设计、环境设计等学科骨干密切协同，他们把学科建设与社会服务通过对现实问题的回应结合了起来。采取了生动活泼的校内校外、课上课下的教学和研究，可谓硕果累累。我为这些同行勤勉的工作感到骄傲！

在当下的室内设计领域，一个突出实用价值的阶段，唯有高等教育机构还在担当着研究和思考的重任。无论是基础研究还是对现实个案展开的深层次研究，都是在学科里进行的，这不仅弥足珍贵，更不可或缺。作为一个以学术研究为己任的社会团体，这种性质的工作才是我们的本色，因此我们必须对他们长期以来的努力给予支持和帮助。

《出师表》在行事的方法上有许多精辟的概述，如"悉以咨之，然后施行，必能裨补阙漏，有所广益"。在室内设计学科领域，研究和实践在时间上的关系也始终处于螺旋上升的结构状态，实践是研究的开始，研究又是大规模实践的准备。我一直认为室内设计有着广阔的未来，它将是延续几千年的建筑设计未来的存在形式，需要"咨诹善道，察纳雅言"对其本体进行更多的分析研究，对它的未来展开更多的讨论和展望。

写于米兰

education. The Innovation Program of Characteristic Education on "interior Design 6+" Joint Graduation Project is a distinct representative of this type, and the dominator of this program is the defender of the era. For a long time, the IID-ASC has been concerned about design education, which has, unexpectedly, become a highland of the industrial hotspots. It's lonely to be in a high position, but sincerity can be seen this way. Over the six years, the backbones in interior design and environment design at the universities involved have worked closely with one another, combining discipline construction with social services by responding to practical problems. The have done lively teaching and research work on and off campus, in and outside class, making innumerable great achievements. I'm proud of their diligence!

In the field of interior design, as for a stage that highlights the practical value, only institutions of higher education remain obligated to research and think. Whether it is basic research or in-depth research on a practical case, it is done in the disciplinary field. This is not only precious, but indispensable. As a public organization devoted to academic research, we should do such work, so we must offer support and help for their long-term efforts.

Memorial to the Throne for an Expedition contains many insightful overviews of matter handling, such as "consult him about all matters, big and small, before acting, and this can necessarily make good defects and gain more benefits". In the field of interior design, the temporal relationship between research and practice is always on the spiral rise. Practice is the start of research, while research is a preparation for massive practice. I always think that interior design has a bright future, and will become an existent form of the architectural design, which has lasted for thousands of years. We need to "seek advice from others and accept admonitions" to make more analyses and studies of interior design, as well as more discussions and prospects on its future.

Su Dan
Written in Milan

前言
Introduction

陈静勇

《绿色营造卷——建筑室内装配化设
计》执行主编

Executive editor of *Green construction
—building indoor assembly design*

中国建筑学会室内设计分会副理事长

Vice Chairman of the Institute of
Interior Design of Architectural Society
of China

北京建筑大学设计艺术研究院　教授

Professor of Beijing University of Civil
Engineering and Architecture Graduate
School of Architecture Design and Art

"适用、经济、绿色、美观"建筑方针下的联合毕业设计

Joint Graduation Design under the Architectural Principle of
Application, Economy, Environmental-friendliness and Beauty

　　大暑节气，多地酷热，暴雨连连。同往年一样，中国建筑学会室内设计分会"室内设计 6+" 2019（第七届）联合毕业设计《绿色营造卷——建筑室内装配化设计》（简称《绿色营造卷》）的编写工作在暑假开工了。这是中国建筑学会"室内设计 6+"联合毕业设计特色教育创新项目实施的第 7 届成果，也是庆祝中国建筑学会室内设计分会成立 30 周年的纪念版。

　　为了转变城市发展方式、塑造城市特色风貌、提升城市环境质量、创新城市管理服务，2016 年 2 月 6 日发布的《中共中央　国务院关于进一步加强城市规划建设管理工作的若干意见》提出了"适用、经济、绿色、美观"的建筑八字方针。在建设创新型国家和人才强国战略的指引下，面向培养新时代设计人才、服务行业发展的新需求，促进"一流大学和一流学科建设"的新目标，2013 年由中国建筑学会室内设计分会（以下简称"室内分会"）主办，由同济大学、哈尔滨工业大学、华南理工大学、西安建筑科技大学、北京建筑大学、南京艺术学院、浙江工业大学组成的"核心高校组"，依托建筑学"室内设计及其理论"学科方向、设计学"环境设计"学科方向的相关专业，广泛联合知名设计企业，共同创立的"室内设计 6+"联合毕业设计活动，在探索室内设计师卓越培养之路上已走过了 7 年，彰显了中国建筑学会室内设计特色教育创新项目平台的内涵和影响力。

　　进入新时代，我国更加重视建筑工业化体系建设，坚持标准化设计、工厂化生产、装配化施工、一体化装修、信息化管理、智能化应用，提高技术水平和工

It is the Great Heat (12th solar term, on July 22, 23, or 24) when extremely hot weather hits various regions and rainstorm lasts for a long time. As every year, the overall edition of "Green construction—Building Indoor Assembly Design" (hereinafter referred to as "Green building") of "the Interior Design 6+" (the 7th) joint graduation design 2019 has been started by the Institute of Interior Design of Architectural Society of China in summer holiday, which is the seventh achievement implemented by Architectural Society of China in the Featured Educational Innovation Project of 'the Interior Design 6+' Joint Graduation Design and the commemorative version celebrating the 30th anniversary of the establishment of the Institute of Interior Design of Architectural Society of China as well.

In order to change the urban development mode, shape the unique urban appearance, improve urban environmental quality and innovate urban management and service, "Opinions of the Central Committee of the CPC and the State Council on Further Strengthening the Management of Urban Planning and Construction" issued on February 6, 2016 summarized the architectural principles with four words- "application, economy, environmental-friendliness, beauty". Guided by the strategy of constructing innovative country and reinvigorating China through human resource development, and facing with new requirements of cultivating design talents of new era and development of service filed and new goal of promoting "the establishment of top universities and top disciplines", "the core university group" held by the Institute of Interior Design of Architectural Society of China (hereinafter referred to as "the Institute of Interior Design") in 2013 and composed of Tongji University, Harbin Institute of Technology, South China University of Technology, Xi'an University of Architecture and Technology, Beijing University of Civil Engineering and Architecture, Nanjing University of the Arts, Zhejiang University of Technology, relying on the "interior design and its theory" of Architecture and "environment design" of Design Science and its relevant majors, united various well-known design enterprises and established "the Interior Design 6+"joint graduation design activity together with 7 years' history on exploring interior designer cultivation, which reflected the connotation and influence of the innovative project platform for interior design feature education of the Architectural Society of China.

On entering a new era, China pays more attention to the construction of industrialized architecture system, adheres to standardized design, factory production, assembly construction, integrated decoration, information management, and intelligent application, improves technical level and engineering quality and promotes transformation and upgrading of the architecture industry. As was noted in the "Guidance of the General Office of State Council on Developing the Assembly Building" （Guidance for Short） issued in 2016, we shall firmly establish

程质量，促进建筑产业转型升级。2016 年发布的《国务院办公厅关于大力发展装配式建筑的指导意见》（以下简称《指导意见》，将发展装配式建筑作为重大发展战略，提出了要"牢固树立和贯彻落实创新、协调、绿色、开放、共享的发展理念，按照适用、经济、安全、绿色、美观的要求，推动建造方式创新，大力发展装配式混凝土建筑和钢结构建筑，在具备条件的地方倡导发展现代木结构建筑，不断提高装配式建筑在新建建筑中的比例。坚持标准化设计、工厂化生产、装配化施工、一体化装修、信息化管理、智能化应用，提高技术水平和工程质量，促进建筑产业转型升级"的发展装配式建筑指导思想。

来自"室内设计 6+"联合毕业设计"核心高校组"建筑学（室内设计）、环境设计、产品设计、艺术与科技等专业的毕业生们，通过学会、高校、企业、专家之间的联合设计教育平台，在"绿色营造——建筑室内装配化设计"总命题下遴选适宜子课题，协同开展了综合性实践教学活动，从多专业领域感受《指导意见》提出的"健全标准规范体系、创新装配式建筑设计、优化部品部件生产、提升装配施工水平、推进建筑全装修、推广绿色建材、推行工程总承包"重点任务，促进了毕业设计教学水平和人才培养质量的提升。

本届联合毕业设计项目主要采取"6+N 环节"的组织模式：

（1）命题研讨。2018 年 10 月 27 日（重庆市，重庆商学院），2018 室内设计第二十八届（重庆）年会"室内设计 6+"2019（第七届）联合毕业设计命题研讨会。

（2）开题调研。2019 年 3 月 2—3 日（北京市，北京建筑大学），开题调研、专题讲坛、项目考察。

（3）中期检查。4 月 20—21 日（杭州市，浙江工业大学），中期检查、项目考察。

（4）毕业答辩。6 月 1—2 日（南京市，南京艺术学院），毕业答辩、项目考察、活动总结。

（5）编辑出版。7 月 21 日—11 月 17 日，室内分会和北京建筑大学负责编写中国建筑学会室内设计分会推荐专业教学参考书：'室内设计 6+'2019（第七届）联合毕业设计《绿色营造卷——建筑室内装配化设计》，并由中国水利水电出版社于 2019 室内分会第二十九届（上海）年会期间出版发行。

（6）专题展览。11 月 15—17 日，"室内设计 6+"2019（第七届）联合毕业设计专题展览在室内分会 2019 二十九届（上海）

and implement the development concept of innovation, coordination, environment-friendliness, openness, sharing with the development of assembly building as a major development strategy, promoting the innovation of construction mode according to requirements of application, economy, security, environmental-friendliness and beauty. The assembly concrete and steel buildings shall be developed with more efforts. Modern wood buildings shall be developed where there are conditions as advocated with increasing proportion of assembly buildings in the newly-build buildings. The guiding concept of standardized design, factory production, assembly construction, integrated decoration, information management, intelligent application, improving the technology level and engineering quality and promoting transformation and upgrading of the architecture industry shall be adhered to on developing assembly buildings.

Graduates majored in Architecture (Interior design), Environment Design, Product Design, Arts and Technology from the "core university group" of "the Interior Design 6+" joint graduation design have selected the appropriate sub-project under the general proposition of "Green construction — building indoor assembly design" through the joint design education platform of the Institute, universities, enterprises and experts, collaboratively carrying out a comprehensive practical teaching activity from which they have understood the key tasks of "perfecting standard system, innovating assembly architectural design, optimizing parts production, promoting assembly construction level, full decoration of buildings, environmental-friendly building materials and overall contracting of engineering" proposed by the "Guidance" in multiple professional fields. In this way, the education level of graduation design and talent cultivation quality has been improved.

The organization pattern of "6+N stage" is adopted in current joint graduation design project:

(1) Proposition research. On October 27, 2018 (in Chongqing, Chongqing Technology and Business University), the 28th (Chongqing) Annual Meeting for Interior Design in 2018, "the Interior Design 6+" (the 7th) Proposition Seminar for Joint Graduation Design in 2019.

(2) Proposal research. from March 2 to March 3, 2019 (in Beijing, Beijing University of Civil Engineering and Architecture), proposal research, thematic forum, project survey.

(3) Mid-term check. from April 20 to April 21 (in Hangzhou, Zhejiang University of Technology), mid-term check, project survey.

(4) Graduation defense. from June 1 to June 2 (in Nanjing, Nanjing University of the Arts), graduation defense,project survey, activity summary.

(5) Edition and publish. From July 21 to November 17, the professional teaching reference books recommended by the Institute of Interior Design of the Architecture Society of China: "Environmental-friendly Construction [volume] -Building Indoor Assembly Design" of "the Interior Design 6+"2019 (the 7th) joint graduation design shall be edited by the Institute of Interior Design and Beijing University of Civil Engineering and Architecture and published and issued by China Water &Power Press during the 29th (Shanghai) annual meeting of 2019 Institute of Interior Design.

(6) Thematic exhibition. From November 15 to November 17, the thematic exhibition for "the Interior Design 6+" (the 7th) joint graduation

年会暨国际学术交流会期间举办。

"N+"指联合拓展。本项目还设置了一些特色拓展环节。在联合毕业设计活动中，室内分会组织参加高校、命题单位、行业专家之间开展对毕业设计的联合指导。聘请了项目观察员作观察点评，提出观察意见与建议，促进项目不断优化。2019年起，室内分会以通过全国建筑学专业评估学校和设计学学科评估高校为骨干和带动，开始逐步建设起"室内设计6+"联合毕业设计活动"地区高校组"（华北、华东、华南、华西、华中、东北）和"省市高校组"，与"核心高校组"协同开展"校组交流"，进一步推进服务设计人才联合培养平台的整体提升工作。《绿色营造卷》主要内容采用中英文对照方式编写，是作为中国建筑学会室内设计分会（IID-ASC）与亚洲室内设计联合会（AIDIA）等相关国际学术组织开展设计教育对外交流的特色案例之一。

《绿色营造卷》编制工作继续优化本项目前6届形成的书籍设计特色。本卷"项目规章""开题调研""中期检查""毕业答辩""教育研究""专题讲坛""风采定格"等章涵盖了"室内设计6+"活动的主要环节，记录了项目主体内容和过程，附有专家点评、学生感言、获奖证书、活动照片等。"项目规章"一章收录项目《章程》《答辩工作细则》《框架任务书》《纲要》等，是有章可循，建立长效机制的体现。

"室内设计6+"2019（第七届）联合毕业设计特色教育创新项目已顺利完成。从联合毕业活动中走来的建筑学（室内设计）、环境设计、产品设计、艺术与科技等专业的毕业生们，获得了相应的建筑学学士（专业学位）、工学学士、艺术学学士学位。继而，他们有的考取了硕士研究生，有的出国留学深造，有的直接就业工作，共同踏上了面向行业需求、走向卓越发展的新征途。

今年是中国建筑学会室内设计分会成立30周年，"室内设计6+"联合毕业设计特色教育创新项目走过了7年的探索历程。在此，以《绿色营造卷》作为纪念版呈献给朋友们。

感谢全国高等学校建筑学学科专业指导委员会、教育部高等学校设计学类专业教学委员会等长期以来对高校相关学科专业建设工作的指导！

感谢相关高校、命题单位、支持单位、讲坛专家、特邀专家、出版单位等对"室内

design 2019 shall be held during the 29th (Shanghai) annual meeting of the Institute of Interior Design and International Conference for Academic Exchange 2019.

"+N" means Joint expansion. There are some special expanding activities set in this project. In the joint graduation design activity, the participating universities, proposition units, experts in this field are organized to conduct "joint instruction" for the graduation design by the Institute of Interior Design, with project inspectors being employed to "inspect and evaluate" and give their opinions and suggestions through inspection so as to continuously optimize the project. Since 2019, with the impetus and the key role of the national architecture evaluation schools and discipline evaluation universities for design, the Institute of Interior Design has gradually established the "regional university group" (in North China, East China, South China, West China,Central China, Northeast China) and "provincial and municipal university group" of the "Interior Design 6+" joint graduation design activity which together with the "core university group" conduct the "university-group communication" , further promoting the overall improvement of joint cultivation platform for design talents. The main content of the "Green construction" is edited with both English and Chinese as one of the characteristic examples for "international communication" of design education organized by the Institute of Interior Design of Architecture Society of China (IID-ASC) and Asia Interior Design Institute Association (AIDIA) and other related international academic organizations.

The preparation of "Green construction" continues to optimize the design feature formed in the previous six copies.The core content of "project regulations", "proposal research", "mid-term inspection", " graduation defense", "educational research" "thematic forum" "achievements records" of this Volume includes main stages of the "Interior Design 6+", recording the major content and process of the project attached with experts evaluation, students speech, rewarded certificates, photos of the activities, etc. The "project regulations" incorporate "Charter", "Detailed Rules of Defense","Framework Task Book","Outline", etc. which is a reflection that there are rules to follow and long-term mechanism to establish.

Featured Educational Innovation Project of "the Interior Design 6+" (the 7th) joint graduation design 2019 has been successfully completed. Graduates majored in Architecture (Interior design), Environment Design, Product Design, Arts and Technology coming out from the joint graduation activity has been conferred with bachelor of Architecture (professional degree), bachelor of engineering, bachelor of the Arts. Later on, some of them are admitted to be postgraduates, some furthering their education overseas, others going to work directly, embarking on a new journey to meet the needs of the industry and to achieve remarkable development.

This year marks the 30th anniversary of the establishment of the Institute of Interior Design of Architecture Society of China. "The Featured Educational Innovation Project of 'the Interior Design 6+' Joint Graduation Design" has gone through 7 years of exploration. The "Green construction" is hereby presented to you as s commemoration.

Thanks are extended to National Professional Directing Committee for

设计 6+" 联合毕业设计项目的支持和帮助！

感谢项目参加高校的鼎力承办，以及志愿者们的辛勤付出！

"室内设计 6+" 联合毕业设计特色教育创新项目将继续贯彻 "适用、经济、绿色、美观" 的建筑方针，更加广泛地联合校内外力量，培养造就创新能力强、服务行业发展需要的高质量室内设计人才。

二〇一九年七月二十三日（己亥·大暑）

Architecture of Universities, Professional Teaching Committee of Ministry of Education for Design of Universities for their long-term guidance on the establishment of the related disciplines of universities!

It is grateful to relevant universities, proposition units, supporting units, experts of the forum, invited experts, punishing units for their support and help on the "Interior Design 6+" joint graduation design project!

We appreciate participating universities for their undertaking with efforts and to volunteers for their hard working.

"The Featured Educational Innovation Project of 'the Interior Design 6+' Joint Graduation Design" shall continue to implement the architecture principles of "application, economy, environmental-friendliness, beauty", unite more internal and external forces of universities and cultivate high-quality interior design talents with strong innovation ability to serve the development needs of the industry.

July 23, 2019

目　　录
Contents

中期检查

答辩评审

目录

教育研究

风采定格

绿色营造卷——建筑室内装配化设计

联合指导　服务需求

热点命题　纷显特色

项目规章

规矩方圆

以章程为准绳 明确项目任务 纲要和细则
指导项目实施 规范书籍设计

「室内设计6+」2019(第七届)联合毕业设计
"Interior Design 6+"2019(Seventh Year) Joint
Graduation Project Event

"室内设计6+"联合毕业设计章程（2019版）

"Interior Design 6+" Joint Graduation Project Event Charter（2019）

为服务城乡建设领域室内设计专门人才培养需求，加强室内设计师培养的针对性，促进相关高等学校在专业教育教学方面的交流，引导面向建筑行（企）业需求开展综合性实践教学工作，由中国建筑学会室内设计分会（以下简称"室内分会"）倡导、主管，国内外设置室内设计相关专业（方向）的高校与行业代表性建筑与室内设计企业开展联合毕业设计。

为使联合毕业设计活动规范、有序，形成活动品牌和特色，室内分会在征求相关高等学校意见和建议的基础上形成原《"室内设计6+1"校企联合毕业设计章程》，于2013年1月13日"室内设计6+1"2013（首届）校企联合毕业设计（北京）命题会上审议通过，公布试行，并结合活动实际持续修订。

历经2013—2017年连续五届联合毕业设计的深入交流，原"室内设计6+1"校企联合毕业设计取得了丰富成果，形成一定影响力，积累了室内分会设计教育平台建设成功经验，形成了多联融合的特色教育创新项目组织实施格局。2017年10月，室内分会第八届理事会通过《教育工作规划纲要（2017—2025年）》，将该活动更名为"室内设计6+"联合毕业设计。2018年该活动经中国建筑学会批准为"'室内设计6+'联合毕业设计特色教育创新项目"。为此，室内分会编制新版章程，并公布试行。

一、联合毕业设计设立的背景、目的和意义

党的十九大报告指出："建设教育强国是中华民族伟大复兴的基础工程，必须把教育事业放在优先位置，加快教育现代化，办好人民满意的教育。""一流大学和一流学科建设"是建设高等教育强国、实现十九大提出的"实现社会主义现代化和中华民族伟大复兴"总任务的必然选择和重要举措。

自1992年5月开始的全国建筑学专业评估全面引导和提升了我国建筑学专业教育水平，同时也带动了室内设计专业（方向）建设和发展，截至2019年5月，通过全国建筑学专业评估的学校已达69所。

2010年教育部启动了"卓越工程师教育培养计划"，并于2011—2013年分三批公布了进入"卓越计划"的本科专业和研究生层次学科。

2011年国务院学位委员会、教育部公布《学位授予和人才培养学科目录（2011年）》，增设了"艺术学（13）"学科门类，将"设

To meet the demand for professional interior design talent training in the urban and rural construction field, strengthen the pertinence of interior designer cultivation, promote related universities to exchange ideas with one another on professional education and teaching, and assist the universities in carrying out comprehensive practice teaching to meet the needs of the construction industry (enterprises), the domestic and foreign universities that offer interior design-related specialties and representative architectural and interior design enterprises cooperate in doing graduation project design on the initiative of Architectural Society of China and Institute of Interior Design (IID-ASC).

In order that the joint graduation project event should go with a swing in an orderly manner, the IID-ASC formulated the original 《Architectural Society of China and Institute of Interior Design (IID-ASC)》"Interior Design 6+1" University-enterprise Joint Graduation Project Event Charter based on opinions and suggestions from the universities involved. Later, the charter was deliberated, approved and published on a trial basis at the topic-assignment meeting of "Interior Design 6+1" 2013 (First Year) University-enterprise Joint Graduation Project Event in Beijing on January 13th, 2013. Then, the continuous revision was carried out based on actual activities.

After in-depth exchange of views on the joint graduation project event in the past 5 years from 2013 to 2017, the original "Interior Design 6+1" University-enterprise Joint Graduation Project Event achieved fruitful results, accumulating successful experience in the construction of an educational platform for interior design. In October 2017, the 8th council of the IID-ASC passed the Educational Planning Framework (2017-2025), renaming it China "Interior Design 6+" Joint Graduation Project. In 2018, the activity was approved by the Architectural Society of China as the Featured Educational Innovation Project of " the Interior Design 6+" Joint Graduation Design. To this end, the IID-ASC prepared a new version of the charter and announced the trial.

I. The Background, Objective and Significance of the Joint Graduation Project Event

The report of the 19th CPC National Congress indicates: "It is a foundation project related to the rejuvenation of China to build China into a great power of education. This necessitates making the educational cause a priority, quickening educational modernization and meeting the people's requirements for education." The "construction of world-class universities and subjects" is an inevitable choice and important measure to build China into a great power of higher education and fulfill the overall task of "realizing socialist modernization and the rejuvenation of China" set at the 19th CPC National Congress.

Since May 1992, the national architecture major assessment has comprehensively guided and promoted the education level of architecture major in China, and also promoted the construction and development of interior design major (direction). By May 2019, 69 schools had passed the national architecture major assessment.

In 2010, the Ministry of Education initiated the "Excellent Engineer Training Program"; in 2011-2013, it made public the undergraduate

计学（1305）"设置为"艺术学"学科门类中的一级学科。"环境设计"建议作为"设计学"一级学科下的二级学科；"室内设计及其理论"建议作为新调整的"建筑学（0813）"一级学科下的二级学科。

2012年教育部公布《普通高等学校本科专业目录（2012年）》，在"艺术学"学科门类下设"设计学类（1305）"专业，"环境设计（130503）"等成为其下核心专业。

"艺术学"门类的独立设置，设计学一级学科以及环境设计、室内设计等学科专业的设置与调整，形成了我国环境设计教育和室内设计专门人才培养学科专业的新格局。

2015年10月，国务院发布《统筹推进世界一流大学和一流学科建设总体方案》。

2017年1月，教育部、财政部和国家发展改革委印发《统筹推进世界一流大学和一流学科建设实施办法（暂行）》。

因此，组织开展室内设计领域联合毕业设计，对加强相关学科专业特色建设，深化综合性实践各教学环节交流，促进室内设计教育教学协同创新，培养服务行（企）业需求的室内设计专门人才，具有十分重要的意义。

二、联合毕业设计组织机构

1. 指导单位和主办单位

"室内设计6+"联合毕业设计由室内分会主办，受全国高等学校建筑学学科专业指导委员会、教育部高等学校设计学类专业教学指导委员会等指导。

2. 参加高校、联合主办高校和总（参）编高校

联合毕业设计一般由学科专业条件相近、设置室内设计方向的相关专业的6所高校间通过协商、组织成为活动参加高校组，并以通过全国建筑学专业评估学校作为核心高校。应突出参加高校组合的地理区域、办学类型、专业特色、就业面向等的代表性、涵盖性、多样性，在学科专业间形成一定的交叉性和联合毕业设计工作环境和交流氛围。

室内分会组织建立"室内设计6+"联合毕业设计特色教育创新项目三个层级的参加高校组，进一步提升"室内设计6+"联合毕业设计既有"核心高校组"对全国活动的引领和示范作用；在室内分会工作六大地区（华北、华东、华南、华西、华中、东北地区）增设"室内设计6+"联合毕业

specialties and postgraduate programs listed in the "Excellent Program" in three times.

In 2011, the Academic Degree Commission of the State Council and the Ministry of Education released the Catalogue of the Degree and Talent Training Subjects (2011), additionally offering the "art (13)" specialty, setting the "design (1305)" as a first-level topic in the "art" specialty. The "environment design" was proposed as a second-level topic subordinate to the "design", and the "interior design and theory" was proposed as a second-level topic subordinate to the "architecture (0813)", which had just been upgraded to be a second-level subject.

In 2012, the Ministry of Education released the Catalogue of the Undergraduate Programs offered at Regular Institutions of Higher Education (2012), putting the "design (1305)" and "environment design(130503)" under the category of the "art" as core specialties. The setup and adjustment of the "art", as well as the design, a first-level subject, and the interior design formed a new pattern of professional interior design talent training in the area of environment design education.

In October 2015, the State Council issued An Overall Plan on Comprehensively promoting the Construction of World-class Universities and Subjects. In January 2017, the Ministry of Education, the Ministry of Finance and the National Development and Reform Commission printed and issued Measures for Comprehensively promoting the Construction of World-class Universities and Subjects (Interim).

Therefore, the joint graduation project of interior design is of great significance to strengthening the characteristic construction of relevant subjects and specialties, deepening exchanges on comprehensive practical teaching, promoting collaborative innovation of interior design education and teaching, and training professional interior design talents required for service industries (enterprises).

II. The Organizer of the Joint Graduation Project Event

1. Guider and organizer

The China "Interior Design 6+" Joint Graduation Project Event is organized by the IID-ASC, and guided by the Professional Guiding Committee for Architecture at Chinese Universities, the Ministry of Education and other various professional design teaching guidance committees.

2. Universities involved, Co-sponsored university, and Chief Compiler (Co-compilers)

The joint graduation project event is generally attended by 6 universities that offer similar subjects and specialties, including the interior design, through consultation. These universities' geographic region, teaching type, professional characteristics and employment orientation should be highlighted, to hold different disciplines together, and create an environment and atmosphere of communication for collaborative design.

To give full play to the role of the Characteristic Educational Innovation Program on China "Interior Design 6+" Joint Graduation Project in "double first-class construction", the China ××(region) "Interior Design 6+" Joint Graduation Project Event and China ××(province/city) "Interior Design 6+" Joint Graduation Project Event can be carried gradually out in the six

设计"××（地区）高校组"，开展地区活动，突出地区特色；在有条件的省市增设"室内设计 6+"联合毕业设计"××（省/市）高校组"，开展省市活动，突出省市特色。室内分会安排专家、评委、项目观察员等指导不同层级参加高校组联合毕业设计活动，促进多联合融合交流。

每年通过各层级活动参加高校组申报和室内分会遴选，确定相应的毕业设计开题调研、中期检查、毕业答辩等集中活动的联合主办高校，以及中国建筑学会室内设计分会推荐专业教学参考书——"室内设计6+"××（年）（第×届）（××地区或××省/市）联合毕业设计《××（主题）卷——××（总命题）》（以下简称《主题卷》）主编高校，其他参加高校作为参编高校。

每所高校参加联合毕业设计到场汇报的学生一般以 6 人为宜，分为 2 个方案设计组；要求配备 1～2 名指导教师，其中至少有 1名指导教师具有高级职称；高校导师熟悉建筑学（室内设计）、环境设计、产品设计、艺术与科技等参加专业的实践业务，与相关领域企业联系较广泛。室内分会负责聘任高校导师，指导开展联合毕业设计。

3. 命题单位

参加高校向室内分会推荐所在地区、省市的行业代表性建筑与室内设计企业作为毕业设计命题单位，单位命题人应具有高级职称；室内分会负责聘任单位命题人作为联合毕业设计特聘导师。特聘导师与相应高校导师联合编制联合毕业设计总命题下的《××（子课题）毕业设计教学任务书》，指导开展联合毕业设计。

4. 支持单位

通过室内分会联系和参加高校推荐等，遴选每届活动支持单位。由室内分会与支持单位商洽签订活动支持与回馈协议，负责聘任支持单位代表为项目观察员，参与联合毕业设计观察点评。

5. 出版单位

室内分会和《主题卷》主编高校遴选行业知名出版单位，作为《主题卷》出版单位，参与联合毕业设计相关环节工作。

三、联合毕业设计流程环节

（1）联合毕业设计每年由室内分会主办 1 届，与参加高校毕业设计教学工作实际相结合。

（2）室内分会负责联合毕业设计总体策划、宣传，组织研讨、编制、公布每届联合

regions of China (North China, East China, South China, West China,Central China and Northeast China) and all provinces and cities respectively in accordance with related representative characteristic universities' conditions and will to combine with one another.

The annual co-sponsored university of opening research, in-process inspection and graduation defense, as well as the chief compiler of the professional teaching reference book recommended by the IID-ASC: China "Interior Design 6+"××(Year) (the xth) Joint Graduation Project ××(Subject) volume—××(General Assignment)(hereinafter referred to as "the Topic Volume"), is selected from the universities involved by the IID-ASC, while the other universities serve as co-compilers.

Each university shall have 6 students attend the joint graduation project event and be present to make a report, and they shall be subdivided into 2 design teams; 1~2 supervisors, including at least one with a senior professional title, take part in the event with the students; university supervisors are familiar with engineering practice, including Architecture (Interior Design), Environment Design, Product Design, Art and Technology and stay in touch with many enterprises concerned. The IID-ASC is obliged to engage supervisors from the universities involved and guiding the development of Joint Graduation Projects.

3. Topic Assigner

Each university involved shall recommend to the IID-ASC a representative regional or provincial /municipal interior design enterprise as the topic assigner, and the enterprise shall have a senior professional title; the IID-ASC shall hire the topic assigner as a distinguished supervisor of the joint graduation project event. Distinguished tutors and corresponding university tutors should jointly compile the ××(sub-topic) Teaching Task Book of Graduation Design under the general proposition of joint graduation design, and guide the development of joint graduation design.

4. Corporate Supporter

The supporting units of each activity should be selected by contacting and participating in the recommendation of colleges and universities through the IID-ASC. The IID-ASC will negotiate and sign the agreement of activity support and feedback with the support unit, and appoint the representative of the support unit as the project observer to participate in the observation and comment of the joint graduation design.

5. Publishing Enterprise

The IID-ASC and the chief compiler of the Topic Volume shall select a well-known publishing enterprise as the Topic Volume and a participant in the joint graduation project event.

III. Joint Graduation Project Process

(1) The join graduation project event, being held once a year by IID-ASC, which is in line with the universities' graduation project teaching.

(2) The IID-ASC should be responsible for the overall planning and publicity of the joint graduation design, organizing seminars, compiling and publishing each joint graduation design ×× (Theme)—×× (Total Proposition) Framework Task Book, Project Outline, etc., coordinating to participate

毕业设计《××（主题）——××（总命题）框架任务书》《项目纲要》等，协调参加高校、命题单位、相关机构等，聘请领域专家为专题论坛演讲人，组织对毕业设计子课题成果、毕业设计组织单位、毕业设计命题单位等的审核，以及室内设计教育国际交流等。

（3）联合毕业设计主要教学环节包括命题研讨、开题调研、中期检查、毕业答辩、编辑出版、专题展览6个主要环节，以及联合指导、观察点评、跨校交流、对外交流等多个联合毕业设计活动的扩展环节。相关工作分别由室内分会、参加高校、命题单位、支持单位、出版单位等分工协同落实。

（4）命题研讨。室内分会组织召开联合毕业设计命题研讨会。每届联合毕业设计的总命题着眼建筑学（室内设计）、环境设计、产品设计、艺术与科技等相关领域学术前沿和行业发展热点问题，参加高校联合命题单位细化总命题下子课题。联合毕业设计子课题要求具备相关设计资料收集、现场踏勘、项目管理方支持等条件。

命题研讨会一般安排在高校秋季学期，在当年室内分会年会期间（10月下旬）安排专题研讨。

（5）开题调研。室内分会组织开展联合毕业设计开题调研活动，颁发联合毕业设计高校导师和特聘导师聘书；联合主办高校协同落实开题仪式、专题论坛、开题报告汇报、项目调研等工作。每所参加高校进行开题报告汇报，合组不超过20分钟，专家点评不超过10分钟。

开题活动一般安排在高校春季学期开学初（3月上旬）进行。

（6）中期检查。室内分会组织开展联合毕业设计中期检查活动；联合主办高校协同落实专题论坛、中期检查汇报、项目调研等工作。每所参加高校推荐不超过2个初步设计方案组进行汇报，每组陈述不超过20分钟，专家点评不超过10分钟。

中期检查一般安排在春季学期期中（4月中旬）进行。

（7）毕业答辩。室内分会组织开展联合毕业设计答辩活动；联合主办高校协同落实毕业答辩、颁发证书、项目调研等工作。每所参加高校推荐不超过2个深化设计方案组进行陈述与答辩，每组陈述不超过20分钟，专家点评与学生回答不超过10分钟。

在答辩、点评的基础上，室内分会组织开展"室内设计6+"联合毕业设计特色教

in colleges and universities, proposition units, related institutions, etc., hiring field experts as special forum speakers, organizing the evaluation of graduation design sub-project results, graduation design organization units, proposition units to graduation design, international exchange of interior design education and so on.

(3) The process of joint graduation project teaching consists of 6 parts: topic assignment discussion, in-process inspection, graduation defense, compilation and publication of the Topic Volume and special exhibition. In addition, joint guidance observation and comment the communication of cross-school and external exchange are the extension of the joint graduation project events. The above work shall be done by the IID-ASC, universities, topic assigners, supporters and publishing enterprises separately or synergistically.

(4) Topic Assignment Discussion. The IID-ASC shall organize a discussion on topic assignment for the joint graduation project event. The general assignment of each year's joint graduation project event shall be focused on the academic frontier and industrial hot spots of Architecture (Interior Design), Environment Design, Product Design, Art and Technology. Besides that, the universities involved shall make a subtopic under the general assignment. The subtopic of the joint graduation project event requires design data acquisition, reconnaissance trip, and support from construction management.

The proposition seminar will be usually arranged in the fall semester of colleges and universities. The special seminar will be arranged during the annual meeting of the Institute of Interior Design (late October).

(5) Opening Research. The IID-ASC should organize the joint graduation design and opening research activities, award appointment letters of the joint graduation design college tutors and special tutors and co-sponsored university's collaborative implementation of the opening ceremony, special forums, opening report reports, project research and other work. Each participating university will report on the opening report, the report combination will not exceed 20 minutes, and the expert review will not exceed 10 minutes.

Opening research activities are usually arranged at the beginning of spring term (early march).

(6) In-process Inspection. The IID-ASC should organize the mid-term inspection activities of the joint graduation design. Co-sponsored university's collaborative implementation of the special forum, the mid-term inspection report, the project research and other work which will be implemented under the cooperation by colleges and universities. Each participating university should recommend no more than 2 preliminary design scheme groups for reporting; wherein, each group makes a presentation for no more than 20 minutes and the expert comments for no more than 10 minutes.

The mid-term inspection is usually arranged in the middle of the spring term (mid-april).

(7) Graduation Defense. The IID-ASC shall organize the joint graduation design defense activity. Co-sponsored universities to coordinate the implementation of graduation defense, certificate issuance,

育创新项目年度研讨，重点研究各毕业设计子课题成果质量，肯定毕业设计组织单位、毕业设计命题单位、支持单位等。坚持"质量第一、宁缺毋滥"的原则，毕业设计子课题成果成绩按百分制计，其中90～100分、80～89分两段打分结果一般按照1:2比例设置。

　　毕业答辩一般安排在春季学期期末（6月上旬）进行。

　　（8）专题展览。室内分会在每届联合毕业设计结束当年的室内分会年会暨学术研讨会（每年10—11月）举办期间安排联合毕业设计作品专题展览；专题展览结束后，相关高校可自愿向室内分会申请联合毕业设计作品巡回展出。

　　（9）编辑出版。基于每届联合毕业设计成果，由室内分会组织编辑出版《主题卷》，作为室内分会推荐的专业教学参考书。《主题卷》的编写工作由室内分会、主编高校和参编高校共同完成。参编高校导师负责本校排版稿的审稿等工作，出版单位负责审校、出版、发行等工作。

　　（10）对外交流。室内分会和出版单位一般在每届联合毕业设计结束当年、室内分会年会期间联合举办《主题卷》发行仪式。室内分会联系如亚洲室内设计联合会（AIDIA）等室内设计国际学术组织，开展室内设计教育成果国际交流，宣传中国室内设计教育，拓展国际交流途径。

　　四、联合毕业设计相关经费

　　（1）室内分会负责筹措对毕业设计项目子课题成果（含完成人、指导教师）、毕业设计组织单位、毕业设计命题单位、支持单位等的专家差旅、劳务经费，以及室内分会年会专题展览、宣传经费和《主题卷》出版补充经费等。

　　（2）参加高校自筹参加联合毕业设计相关师生各环节经费。

　　（3）联合主办高校负责联合毕业设计开题调研、中期检查、毕业答辩与颁发证书等环节的宣传、场地、设备、调研等经费；毕业答辩与颁发证书环节，联合主办高校还负责用作毕业答辩的深化设计方案《主题卷》书稿册页的打印装订等经费；《主题卷》主编高校负责出版主体经费等，并为项目成果交流提供样书。

　　（4）命题单位、支持单位、出版单位等负责为向校企联合毕业设计提供一定形式的支持等。

project research and other work. Every university involved shall choose no more than 2 further detailed design of schema group for presentation and oral defense. To be specific, either team shall give a presentation of up to 20min; expert comments and student answers shall take up to 10min.

On the basis of the defense and comments, the Institute of Interior Design should organize the annual discussion with respect to the Featured Educational Innovation Project of " the Interior Design 6+" Joint Graduation Design, with the focus on evaluating the quality of the sub-projects of graduation designs, and evaluating the organization units of graduation designs and the proposition units, the support units of graduation designs and so on. Under the principle of " quality First, put quality before quantity", the results of graduation design sub-subjects are calculated based on the percentage system, among which the evaluation results of 90-100 points and 80-89 points are generally set at 1:2 ratio.

Generally, graduation defense is performed in the late spring term (early June).

(8) Special Exhibition. The IID-ASC organizes a special exhibition of joint graduation project contents after the joint graduation project event ends and during the period of the annual meeting and academic conference; after the end of the special exhibition, the universities concerned can voluntarily apply to the IID-ASC for an itinerant exhibition of joint graduation project contents.

(9) Compilation and Publication. The IID-ASC organizes the compilation and publication of the Topic Volume based on each year's joint graduation project results as a professional teaching reference book. The Topic Volume shall be co-compiled by the IID-ASC, the chief compiler and the other universities involved, and each university's supervisors shall be responsible for examining its own manuscript, while the publishing enterprise shall serve as an editor in charge for proofreading and publication.

(10) External Exchanges. Generally, the IID-ASC and publishing enterprise co-distribute the Topic Volume after the end of the joint graduation project event and during the IID-ASC's annual meeting; the IID-ASC shall communicate with the Asia Interior Design Institute Association (AIDIA) and other international academic organizations for interior design on the results of interior design education, propagandize Chinese interior design education and expand the way of international communication.

IV. Costs of Joint Graduation Project

(1) The IID-ASC is responsible for raises funds for evaluation on the quality of the graduation sub-project results excellent graduation project participants and the proposition units, the support units to the graduation project, as well as for the special exhibition propaganda and the supplementary funds of Topic Volume for publication.

(2) Costs of universities' participation in the joint graduation project event.

(3) Co-sponsored universities should be responsible for publicity, venue, equipment, research and other funds of the proposal research for joint graduation design, mid-term inspection, graduation defense, issue the certificate; the university undertaking the graduation defense and review,

（5）室内分会适时组织参加高校组，将"室内设计 6+"联合毕业设计特色教育创新项目申报为国家有关基金项目。

五、附则

本章程于 2019 年 3 月 2 日"室内设计 6+" 2019（第七届）联合毕业设计开题日公布试行，由中国建筑学会室内设计分会负责解释。先前版本废止。

commendation and awards should also assume the fees for printing and binding manuscript pages of the deepened design scheme Theme Volume for graduation defense. The university as the chief editor of Theme Volume is responsible for the main publishing fund, etc., and should provide sample books for the exchange of project achievements.

(4) The topic assigner, supporter and publishing enterprise shall offer some support to the joint graduation project event.

(5) The IID-ASC timely should organize the participating college group to declare the Featured Educational Innovation Project of " the Interior Design 6+" Joint Graduation Design as the relevant national fund project.

V. Supplementary Provisions

This charter will be published on the opening day of 2019 (7th session) joint graduation project of "Interior Design 6+" on March 2, 2019 for trial implementation, with the previous version abolished, and the Institute of Interior Design of the Architectural Society of China will be responsible for the interpretation.

项目规章

"室内设计6+"联合毕业设计特色教育创新项目毕业答辩工作细则（2019版）

Rules for Graduation Defense of Featured Educational Innovation Project of "the Interior Design 6+" Joint Graduation Design (2019)

中国建筑学会室内设计分会（以下简称"室内分会"）依据《"室内设计6+"联合毕业设计特色教育创新项目章程》，制订《"室内设计6+"联合毕业设计特色教育创新项目毕业答辩工作细则》，指导相关单位和人员开展联合毕业设计答辩等工作。

一、答辩准备

（1）参加高校，每校推选指导教师不超过2名。

（2）参加高校，每校推选不超过2个方案深化设计组参加毕业答辩，学生总数不超过6名，完成《主题卷》书稿、答辩汇报PPT、年会展版编制。

（3）每个深化设计方案组按《主题卷》书稿要求准备毕业答辩册页（含中期检查、毕业答辩两阶段成果）、答辩汇报PPT等电子文档，须于毕业设计答辩前1周发送到项目组委会指定邮箱；由承办高校负责汇总打印、装订等，作为毕业答辩材料。

（4）参加高校按《"室内设计6+"联合毕业设计专辑排版要求》编辑本届《主题卷》书稿，须于毕业设计答辩结束后2周内发送到室内分会指定邮箱。

（5）每个深化设计方案编制3张展板，使用室内分会统一发布的模板编辑，展板幅面为A0加长（900mm×1800mm），分辨率不小于100dpi。展板电子版须于联合毕业设计当年室内分会年会前1月发送到年会组委会指定邮箱。室内分会负责打印、布展等。

二、毕业答辩

1. 毕业设计答辩委员会

毕业设计答辩委员会由室内分会特邀导师和高校导师组成。

（1）室内分会特邀导师委员一般由室内分会特邀专家、命题单位、项目观察员、支持单位代表等在内的7～9名专家担任；答辩委员会组长一般由室内分会提名人选，经答辩委员会集体确认后担任，并主持毕业答辩工作。

（2）高校导师委员由各参加高校分别推选1位本届毕业设计指导教师担任。

2. 毕业设计答辩

（1）毕业设计答辩按"室内设计6+"联合毕业设计教育创新项目各子课题进行打分，成绩按百分制计，其中90～100分、80～89分两段的比例一

The Institute of Interior Design of the Architectural Society of China (hereinafter referred to as "the Institute of Interior Design") has formulated the "Rules for Graduation Defense of Featured Educational Innovation Project of 'the Interior Design 6+' Joint Graduation Design" in accordance with the Project Charter of the Featured Educational Innovation Project of "the Interior Design 6+" Joint Graduation Design, guiding the relevant units and personnel to conduct the reply of the joint graduation design.

I. Preparation for the Reply

(1) No more than 2 tutors shall be selected for each participating university.

(2) Each participating university shall select no more than 2 teams of deepened design scheme to participate in the graduation defense and no more than 6 students to complete preparation of the manuscript pages of Theme Volume, presentation PPT for the reply, exhibition boards for the annual meeting.

(3) Electronic documents such as pages for graduation defense (including achievements in mid-term inspection and graduation defense), presentation PPT for the reply shall be prepared by each team of deepened design scheme according to requirements of the manuscript pages of Theme Volume and delivered to the mailbox designated by the project organizing committee 1 week prior to the graduation design reply which shall be collected, printed and bound by the undertaking university as materials for graduation defense.

(4) Current manuscript pages of Theme Volume shall be edited by participant universities in accordance with Typesetting Requirements for 'the Interior Design 6+' Joint Graduation Design and delivered to the mailbox designated by the Institute of Interior Design within 2 weeks after the completion of graduation design reply.

(5) Three exhibition boards shall be prepared for each deepened design scheme and edited according to the unified format published by the Institute of Interior Design with width of them being extended to: 900mm×1800mm on the basis of A0, the resolution is not less than 100dpi. Electronic version of the boards shall be delivered to the mailbox designated by the organizing committee of annual meeting 1 month prior to the annual meeting of the Institute of Interior Design in the year of joint graduation design which shall be printed and arranged by the Institute of Interior Design.

II. Graduation Defense

1. Committee for graduation design reply.

Committee for graduation design reply is composed of judges specially invited by the Institute of Interior Design and tutors of universities.

(1) The judges invited by the Institute of Interior Design generally consist of 7 to 9 experts including invited experts of the Institute of Interior Design, experts in proposition units, project inspectors, representatives of the supporting unit, etc.; the leader of review committee is generally nominated by the Institute of Interior Design and acts as the leader to preside over the reply after collective confirmation of the judges in the review committee.

(2) One adviser for the current graduation design shall be selected by each participating university to be the university judge in the review committee.

2. The graduation design reply

般为 1:2。

（2）第一轮成绩。参加高校每个毕业设计答辩组的陈述时间不超过 20 分钟，问答（一般安排室内分会特邀导师、高校导师委员各 1 人）不超过 10 分钟。由本届室内分会特邀导师、高校导师委员共同填写第一轮成绩单，进行排序（如，1 为建议排序第一，2 为建议排序第二，以此类推）。

（3）第二轮成绩。高校导师须回避第二轮打分。由室内分会特邀导师委员以第一轮成绩单为基础，对照各组答辩方案和答辩表现等进行综合审议；由特邀导师委员填写第二轮成绩单，进行排序（如，1 为建议排序第一，2 为建议排序第二，以此类推）。项目组委会负责计票、监票，形成最终成绩单。

（4）室内分会依据最终成绩单拟定课题成绩决议，由室内分会特邀导师委员集体签字。过程成绩单、统计结果、课题成绩决议等由室内分会负责存档。

3. 毕业设计组织单位

联合主办联合毕业设计项目开题调研、中期检查、毕业答辩。《主题卷》主编高校为毕业设计组织单位。

4. 毕业设计命题单位、支持单位

负责项目命题、给予项目支持的单位为联合毕业设计命题单位、支持单位。

三、证书颁发

（1）室内分会组织证书颁发。

（2）由室内分会特邀嘉宾、命题单位代表、支持单位代表、项目观察员、高校代表等颁发"指导教师"证书。

（3）由室内分会为联合主办高校颁发项目"组织单位"证书；为命题单位、支持单位颁发证书。

（4）由室内分会为参加"室内设计 6+"项目的学生颁发"课题结题"证书。证书上印有本届毕业答辩委员的签名，以示纪念。

（5）证书加盖中国建筑学会章。

四、附则

本细则由中国建筑学会室内设计分会负责解释。

(1) The graduation design reply shall be scored in accordance with sub-projects of the Featured Educational Innovation Project of 'the Interior Design 6+' Joint Graduation Design with results being calculated by the percentage system. The proportion of 90-100 points and 80-89 points is generally controlled to 1:2.

(2) The first round results. Each graduation design reply team from participant universities shall complete statement within 20 minutes and question and answers (generally one invited tutor of the Institute of Interior Design and one university tutor) within 10 minutes. The first round of the transcripts shall be ranked (e.g.1 means first recommendation ranking, 2 means second recommendation ranking, and so on) by the invited judges of the Institute of Interior Design together with the university tutors.

(3) Second round results. University tutors shall withdraw from the second round grade stage. The invited tutor of the Institute of Interior Design shall, on the basis of the first round of the transcript, conduct comprehensive review by comparing reply scheme and reply performance of each team, filling in and ranking the second round the transcripts (e.g.1 means first recommendation ranking, 2 means second recommendation ranking, and so on). Project organizing committee shall take charge of ballot counting, ballot supervising and figuring out the result of final the transcripts.

(4)Based on the final the transcripts, the Institute of Interior Design shall formulate the project achievement resolution which shall be signed by every tutor of the Institute. The Institute of Interior Design shall be responsible for archiving the transcripts, counting results and project achievement resolution during the process.

3. Organizing unit for graduation design

University conducting proposal research, mid-term inspection, graduation defense, summarization and publish of current joint graduation design project.

4. Graduation design proposition unit, support unit

Proposition supporting units in charge shall be the joint graduation design proposition unit, support unit.

III. Issue the certificate

(1) Issue the certificate is organized by the Institute of Interior Design.

(2) Special guests, representatives of proposition units, supporting units, project observers, representatives of colleges and universities will be awarded the "instructor" certificate.

(3) The indoor branch shall issue the project "organizational unit" certificate for the co-host university; Issue certificates to propositional units and supporting units.

(4) The indoor club will issue the "project conclusion" certificate to the students who participate in the "6+" project; The certificate is printed with the signature of the graduating defense committee member of the project to commemorate.

(5) The certificate shall be affixed with the seal of China architecture association.

IV. Supplementary Rules

The Institute of Interior Design of Architectural Society of China shall reserve the right to interpret the Rules.

"室内设计 6+" 2019（第七届）联合毕业设计框架任务书

Framework Task Book of (the 7th) "The Interior Design 6+" Joint Graduation Design 2019

中国建筑学会室内设计分会（以下简称"室内分会"）《"室内设计 6+" 2019（第七届）联合毕业设计框架任务书》（简称《2019框架任务书》）是依据《"室内设计 6+"联合毕业设计特色教育创新项目章程》和2018室内设计第二十九届（重庆）年会命题研讨会意见，由"室内设计 6+" 2019（第七届）联合毕业设计"核心高校组"联合命题单位编制形成。参加高校依据《2019框架任务书》，结合本校毕业设计教学工作实际，进一步编制本校《"室内设计 6+" 2019（第七届）联合毕业设计详细任务书》（简称《2019详细任务书》），指导联合毕业设计教学工作。

一、总命题

装配式建筑是用预制部件在工地装配而成的建筑。发展装配式建筑是建造方式的重大变革，是我国当前推进供给侧结构性改革和新型城镇化发展的重要举措，有利于节约资源能源、减少施工污染、提升劳动生产效率和质量安全水平，有利于促进建筑业与信息息化工业化深度融合、培育新产业新动能、推动化解过剩产能。近年来，我国积极探索发展装配式建筑，但建造方式大多仍以现场浇筑为主，装配式建筑比例和规模化程度较低，与发展绿色建筑的有关要求以及先进建造方式相比还有很大差距。

改革开放以来，我国的室内设计与装修行业迅猛发展，规模空前，同时也进行了局域的标准化设计、工厂化生产、装配化施工、一体化装修、信息化管理、智能化应用，提高技术水平和工程质量，促进建筑产业转型升级的尝试和探索，但建筑与室内装配化应用程度和绿色建筑评价程度还没有得到整体提升，在设计专门人才培养上也需要积极适应服务需求，深化专业教育教学改革。为此，室内分会将"绿色营造——建筑室内装配化设计"作为"室内设计 6+" 2019（第七届）联合毕业设计的总命题。参加高校子课题名称在本校《2019详细任务书》中拟定。

在我国的社会主义建设事业中，建筑工业担负着重大的基本建设任务。从第一个五年计划开始，建筑业全体职工积极努力，同时获得各方面的大力支援。但是那一时期的我国建筑工业基础差、技术装备落后，在组织领导和管理制度方面也还存在着很多问题，远不能满足巨大的基本建设任务对建筑业的要求。为了从根本上改善我国的建筑工业，积极地、步骤地实行工厂化、机械化施工，逐步完成对建筑工业的技术改造，逐步完成向建筑工业化的过渡，采用工业化的建筑方

Framework Task Book of (the 7th) "The Interior Design 6+" Joint Graduation Design 2019 ("2019 Framework Task Book") of the Institute of Interior Design of the Architectural Society of China (hereinafter referred to as "the Institute of Interior Design") is prepared by the joint proposition organization of "core college group" of (the 7th) "The Interior Design 6+" Joint Graduation Design 2019 according to Project Charter of the Featured Educational Innovation Project of "the Interior Design 6+" Joint Graduation Design and the comments of the proposition seminar of the 29th (Chongqing) Annual Meeting of Interior Design 2018. The participating colleges shall further prepare Detailed Task Book of (the 7th) "The Interior Design 6+" Joint Graduation Design 2019 ("2019 Detailed Task Book") respectively to guide teaching work of joint graduation design through combining the actual teaching work condition of the college graduation design according to 2019 Framework Task Book.

I. Total Proposition

Assembly building is the building constructed through assembling the prefabricated parts at construction site. The development of fabricated building is significant reform of construction method, and important measure to promote the supply-side structural reform and new urbanization development currently in China. It will help to save resources and energy, reduce construction pollution, improve labor productivity and quality safety level, promote the in-depth integration of construction industry with informationization and industrialization, nurture new industry and new power energy, and drive the dissolve excess capacity. In recent years, China has proactively explored assembly building. However, the construction method is still mainly site casting, and the ratio and scale level of assembly building is low, far behind relevant requirements of green building development and advanced construction method.

Since the reform and opening up, China interior design and decoration industry has developed rapidly into an unprecedented scale, and carried out local standardization design, factory manufacturing, assembly construction, integrated decoration, information management, and intelligent application simultaneously to improve technical level and engineering quality, and promote the try and exploration of construction industry transformation and upgrade. However, the construction and indoor assembly application level and green building evaluation level have not been comprehensively improved. In regard to specialized design talent training, the service demands should be proactively met so as to deepen specialized education and teaching reform. Therefore, the Institute of Interior Design will confirm "Green construction – building indoor assembly design" as the total proposition of (the 7th) "The Interior Design 6+" Joint Graduation Design 2019. The sub-topics of the participating colleges will be determined respectively in 2019 Detailed Task Book.

In the socialist construction cause in China, construction industry has undertaken important basic construction task. Since the first five-year plan, all the employees and workers in construction industry have provided proactive efforts and powerful support in various areas. However, due to poor foundation, backwards technical equipment, and many problems in organization, leadership and management regulations, the construction industry during that period was far from

法，加快建设速度，降低工程造价，保证工程质量和安全施工，1956年国务院发布了《关于加强和发展建筑工业的决定》。

进入新时代，我国更加重视建筑工业化体系建设，坚持标准化设计、工厂化生产、装配化施工、一体化装修、信息化管理、智能化应用，提高技术水平和工程质量，促进建筑产业转型升级。2016年发布的《国务院办公厅关于大力发展装配式建筑的指导意见》（国办发〔2016〕71号），将发展装配式建筑作为重大发展战略，提出了要"牢固树立和贯彻落实创新、协调、绿色、开放、共享的发展理念，按照适用、经济、安全、绿色、美观的要求，推动建造方式创新，大力发展装配式混凝土建筑和钢结构建筑，在具备条件的地方倡导发展现代木结构建筑，不断提高装配式建筑在新建建筑中的比例。坚持标准化设计、工厂化生产、装配化施工、一体化装修、信息化管理、智能化应用，提高技术水平和工程质量，促进建筑产业转型升级"的发展装配式建筑指导思想。《指导意见》提出了"健全标准规范体系、创新装配式建筑设计、优化部品部件生产、提升装配施工水平、推进建筑全装修、推广绿色建材、推行工程总承包"的重点任务，并将"在中国人居环境奖评选、国家生态园林城市评估、绿色建筑评价等工作中增加装配式建筑方面的指标要求"作为加大政策支持的保障措施之一。

二、总体原则

（1）创新装配式建筑设计。统筹建筑结构、机电设备、部品部件、装配施工、装饰装修，推行装配式建筑一体化集成设计。推广通用化、模数化、标准化设计方式，积极应用建筑信息模型技术，提高建筑领域各专业协同设计能力，加强对装配式建筑建设全过程的指导和服务。

（2）优化部品部件生产。引导建筑行业部品部件生产企业合理布局，提高产业聚集度，培育一批技术先进、专业配套、管理规范的骨干企业和生产基地。支持部品部件生产企业完善产品品种和规格，促进专业化、标准化、规模化、信息化生产，优化物流管理，合理组织配送。

（3）提升装配施工水平。引导企业研发应用与装配式施工相适应的技术、设备和机具，提高部品部件的装配施工连接质量和建筑安全性能。鼓励企业创新施工组织方式，推行绿色施工，应用结构工程与分部分项工程协同施工新模式。

meeting the requirements of huge capital construction task. To improve construction industry in China substantially, it proactively undertook factory and mechanized construction step by step, gradually finished the technical transformation of construction industry, and gradually completed the transition to construction industrialization. It adopted industrialized construction method, speeded up construction speed, reduced engineering cost, and guaranteed engineering quality and safe construction. In 1956, the Decision on Strengthening and Developing Construction Industry from the State Council was issued.

In the new era, China attaches more importance to the construction of construction industrialization system, adheres to standardized design, factory manufacturing, and intelligent application, improves technical level and engineering quality, and promotes the transformation and upgrade of construction industry. In 2016, Guiding Opinions of the General Office of the State Council on Vigorously Developing Assembly Building (GBF [2016] No.71) was issued to set the important development strategy of developing assembly building, propose the guiding concept of assembly building development of "firmly establishing and executing the development concepts of innovation, coordination, green, openness and sharing, promoting construction method innovation according to the requirements of application, economical efficiency, safety, green and beauty, vigorously developing assembly concrete building and steel structure building, advocating the development of building in modern wooden structure at the areas where the conditions are met, and continuously increasing the ratio of assembly building in newly-constructed building, adhering to standardized design, factory manufacturing, assembly construction, integrated decoration, information management, and intelligent application, improving technical level and engineering quality, and promoting the transformation and upgrade of construction industry". Guiding Opinions proposed the major task of "improve standard regulation system, innovate assembly construction design, optimize component and part manufacturing, improve assembling level, promote whole decoration of building, publicize green construction materials, and promote project general contracting", and took "add index requirement of assembly building in China habitat environment prize appraisal, national ecological garden city evaluation, and green building evaluation, etc." for one of the assurance measures to enhance policy support.

II. General Principles

(1) Innovate assembly building design. Comprehensively organize building structure, electromechanical equipment, components and parts, assembling, decoration and finishing, and promote integrated design of assembly building. Publicize universal, modularized and standardized design methods, proactively apply building information model technology, improve coordinated design capacity of various majors in construction field, and enhance the whole-process guidance and service for assembly building construction.

(2) Optimize component and part manufacturing. Guide the component and part manufacturers in construction industry to reasonably arrange, increase industry aggregation level, and develop a batch of backbone enterprises and manufacturing bases with advanced technology, professional supporting facilities and standard management. Support the component and part manufacturers to improve product

011

（4）推进建筑全装修。实行装配式建筑装饰装修与主体结构、机电设备协同施工。积极推广标准化、集成化、模块化的装修模式，促进整体厨卫、轻质隔墙等材料、产品和设备管线集成化技术的应用，提高装配化装修水平。倡导菜单式全装修，满足消费者个性化需求。

（5）推广绿色建材。提高绿色建材在装配式建筑中的应用比例。开发应用品质优良、节能环保、功能良好的新型建筑材料，并加快推进绿色建材评价。

（6）推行工程总承包。装配式建筑原则上应采用工程总承包模式，可按照技术复杂类工程项目招投标。工程总承包企业要对工程质量、安全、进度、造价负总责。

（7）确保工程质量安全。完善装配式建筑工程质量安全管理制度，健全质量安全责任体系，落实各方主体质量安全责任。

三、设计范围

依托命题单位提供的新建装配化建筑或既有建筑（可以是非装配化建筑）的室内装配化设计工程项目，整理形成本届联合毕业设计各子课题的设计范围。

四、项目地点

各子课题相应建筑室内装配化设计工程项目地点。

五、设计内容

毕业设计应基于《2019框架任务书》总命题、总体原则，体现绿色营造发展目标下的建筑室内装配化设计专业内容。各校在《2019详细任务书》中明确具体设计内容和要求。

六、主要阶段

（1）开题调研。建筑与室内装配化设计专题调研报告、《绿色营造卷——建筑室内装配化设计》开题调研内容排版页。

（2）中期检查。初步设计方案、《绿色营造卷——建筑室内装配化设计》中期检查内容排版页。

（3）毕业答辩。深化设计方案、《绿色营造卷——建筑室内装配化设计》毕业答辩内容排版页、年会展板。

七、设计成果

（一）设计说明

设计说明内容主要包含：建筑室内装配化设计理念、定位、方案设计分析、经济技术指标等图示及图表等。

varieties and specifications, promote professional, standard, scale and informatized manufacturing, optimize logistics management, and reasonably organize the delivery.

(3) Improve assembling level. Guide the enterprises to research and develop the technology, equipment, machinery and tools applicable to assembly construction, and improve the connection quality of assembly construction of components and parts and building safety performance. Encourage the enterprises to innovate construction organization methods, promote green construction, and apply new model of coordinated construction of structural works and divided and itemized works.

(4) Promote whole decoration of building. Execute coordinated construction of decoration and finishing of assembly building and main structure & electromechanical equipment. Proactively publicize standard, integrated and modularized decoration mode, promote the application of integration technology of materials including integrated kitchen and bathroom and light partition wall, products and equipment pipelines, and improve assembly decoration level. Advocate menu-type whole decoration, and meet the personalized demand of the consumers.

(5) Publicize green construction materials. Increase the application ratio of green construction materials in assembly buildings. Develop new energy-saving and environment-friendly construction materials in good quality with good functions, and speed up the green construction material assessment.

(6) Promote project general contracting. The assembly building should be tendered and bid as engineering project with complicated technology in project general contracting model.

(7) Guarantee engineering quality and safety. Improve quality and safety management regulations of assembly buildings, improve quality and safety responsibility system, and fulfill the quality and safety responsibilities of the entities of all the parties.

III.Scope of Design

The design scope of the sub-topics of the Joint Graduation Design will be formed through sorting the interior assembly design engineering project of newly-constructed assembly building or existing building (non-assembly building is acceptable) provided by the proposition units.

IV. Project Place

The location of the corresponding interior assembly design engineering projects of the sub-topics.

V. Design Contents

The graduation design should be made based on total proposition and general principles of 2019 Framework Task Book, and represent the contents of building interior assembly design major for the purpose of green development. The colleges shall specify the detailed design contents and requirements in 2019 Detailed Task Book.

VI. Main Stages

(1) Opening research: Specialized research report of building and indoor assembly design, and edited page of opening research contents of Green Construction — building indoor assembly design.

(2) Mid-term inspection: Preliminary design scheme and edited page of mid-term inspection of Green Construction — building indoor assembly design.

（二）工程图、分析图表

（1）建筑区域位置图。

（2）建筑及场地总平面图。

（3）建筑平面图、立面图、剖面图、代表性详图、节点、分析图。

（4）室内装配化设计空间的平面图、顶面图、剖（立）视图、代表性详图、节点、分析图。

（5）室内界面设计代表性部品设计图、分析图。

（三）效果图

（1）室内装配化设计空间环境效果图。

（2）室内界面设计代表性部品效果图。

（四）成果提交

1．开题调研

每所参加高校按1个开题调研文件，提交《绿色营造卷——建筑室内装配化设计》开题调研内容排版页、开题调研成果PPT等。

2．中期检查

每所参加高校优选2个初步设计方案，提交《绿色营造卷——建筑室内装配化设计》中期检查内容排版页、中期检查成果PPT等。

3．毕业答辩

（1）每所参加高校优选2个深化设计方案，提交《绿色营造卷——建筑室内装配化设计》毕业答辩内容排版页、毕业答辩成果PPT等。年会展板于答辩评审后单独提交。

（2）每个深化设计方案的展板限3张，展板规格为幅面A0加长（900mm×1800mm）竖版，分辨率不小于100dpi。展板模板由室内分会按照年会展板要求统一提供。年会展览由室内分会负责展板打印、布置。

4．编辑素材

为编辑出版好中国建筑学会室内设计分会推荐专业教学参考书——"室内设计6+"2019（第七届）联合毕业设计《绿色营造卷——建筑室内装配化设计》，相关参加单位和个人等应积极响应室内分会相关工作要求：

（1）单位简介（参加高校、命题单位、支持单位各1篇，中文1000字以内，中英文对照；单位标识矢量文件）。

（2）教学研究论文（每所参加高校导师联名1篇，中文2000～3000字，中英文对照）。

（3）开题调研书稿（每所高校开题调研内容排版占2页或4页，主要标题和关键词等中英文对照）。

（4）中期检查书稿（每所高校2个初步设计方案内容排版各占2页或4页，主要

（3) Graduation defense: Deepening design scheme, edited page of graduation defense contents of Green Construction Volume — building indoor assembly design, and annual meeting panel.

VII. Design Result

(I) Design Explanation

The contents of design explanation mainly include: Concept and positioning of building interior assembly design, scheme design analysis, and drawings and charts of economic and technical indexes, etc.

(II) Engineering Drawing and Analysis Charts

(1) Location map of construction area.

(2) General plan of building and site.

(3) Building floor plan, elevation, profile map, representative detailed drawing, node and analysis chart.

(4) Floor plan, top surface view, profile map (elevation), representative detailed drawing, node and analysis chart of interior assembly design space.

(5) Design chart and analysis chart of representative components and parts of interior interface design.

(III) Rendering

1. Spatial environment rendering of interior assembly design.

2. Representative component and part rendering of interior interface design.

(IV) Result Submission

1. Opening research

Each participating college shall submit edited page of opening research contents of Green Construction — building indoor assembly design and opening research result PPT, and so on for each opening research document.

2. Mid-term inspection

Each participating college shall select 2 preliminary design schemes and submit edited page of mid-term inspection contents of Green Construction — building indoor assembly design, mid-term inspection result PPT and so on.

3. Graduation defense

(1) Each participating college shall select 2 deepening design schemes, and submit edited page of graduation defense contents of Green Construction — building indoor assembly design, and graduation defense result PPT, and so on. The annual meeting panel shall be separately submitted after defense review;

(2) At most 3 panels shall be submitted for each deepening design schemes in the specifications of A0 extended: 900mm×1800mm, in portrait orientation, resolution no less than 100dpi. The panel template shall be unitarily provided by the Institute of Interior Design according to the annual meeting panel requirements. The panels for annual meeting exhibition shall be printed and arranged by the Institute of Interior Design.

4. Edition Materials

To edit and publish professional teaching reference book recommended by the Institute of Interior Design of the Architectural Society of China: (the 7th) "The Interior Design 6+" Joint Graduation Design 2019 Green Construction — Building Indoor Assembly Design, relevant participating organizations, individuals and other shall proactively respond to relevant work requirements of the Institute of

标题和关键词等中英文对照）。

（5）答辩作品书稿（每所高校2个深化设计方案内容排版共占6页或8页，主要标题和关键词等中英文对照）。

（6）演讲提要（每位"专题讲坛"演讲专家1篇，中文800～1000字，中英文对照）。

（7）专家点评（每个深化设计方案分别相应校内外专家、导师点评各1段，中文200～300字，中英文对照）。

（8）学生感言（每个深化设计方案1段，中文200字以内，中英文对照）。

（9）组长总结（毕业答辩委员会组长对本届联合毕业设计答辩总结，中文800～1000字，中英文对照）。

（10）工作照片（每位专家、导师、学生各1张）。

（11）评审证书（室内分会提供，证书电子版）。

（12）活动照片（参加高校提供，各主要环节照片电子版）。

（13）答辩PPT（每所高校2个深化设计方案汇报PPT文件）。

（14）作品展板（每所高校2个深化设计方案展板TIF格式源文件）。

八、附建筑与场地图

见各校《2019详细任务书》。

九、推荐参考选题

（1）"建设可再生的未来"雄安设计中心建筑与环境设计（雄安新区）。

（2）装配式住宅建筑室内设计（上海市宝山区）。

（3）阿里巴巴集团青年公寓室内设计（杭州市余杭区）。

（4）SOHO青年公寓室内设计（南京市江北新区）。

（5）"阿尔山房"民宿示范区建筑与环境设计（内蒙古自治区阿尔山景区）。

（6）华南理工大学教学楼改造设计（广州市大学城）。

（7）中国国际进口博览会展示设计（上海市青浦区）。

Interior Design:

(1) Organization profile (respectively 1 for participating college, proposition organization and supporting organization, within 1,000 Chinese characters, in Chinese and English; vector file for organization logo).

(2) Teaching research thesis (1 joint paper for each tutor of participating college, within 2,000-3,000 Chinese characters, in Chinese and English).

(3) Opening research manuscript (2P or 4P for edited opening research contents of each college, with main titles and key words in Chinese and English).

(4) Mid-term inspection manuscript (respectively 2P or 4P for edited contents of 2 preliminary design schemes of each college, with main titles and key words in Chinese and English).

(5) Defense work manuscript (6P or 8P for edited contents of 2 deepening design schemes of each college in total, with main titles and key words in Chinese and English).

(6) Speech summary (1 summary for each addressing expert for "themed forum", with 800-1,000 Chinese words, in Chinese and English).

(7) Expert comments (respectively 1-paragraph comment for corresponding college internal and external experts and tutors for each deepening design scheme, with 200-300 Chinese characters, in Chinese and English).

(8) Student testimonial (1 paragraph for each deepening design scheme, within 200 Chinese characters, in Chinese and English).

(9) Team leader summary (graduation defense committee leader's comments and summary of the joint graduation design defense, with 800-1,000 Chinese characters, in Chinese and English).

(10) Work photo (1 for each expert, tutor and student).

(11) Assessment certificate (provided by the Institute of Interior Design, certificate in electronic version).

(12) Activity photo (provided by the participating colleges, photos of the main procedures in electronic version).

(13) Defense PPT (report PPT files of 2 deepening design schemes of each college).

(14) Works panel (TIF file of the panels for 2 deepening design schemes for each college).

VIII. Attached Building and Site Drawings

Refer to 2019 Detailed Task Book of each college.

IX. Recommended Topics

(1) "Building a renewable future" – architectural and environment design of Xiong'an Design Center (Xiong'an New Area).

(2) Interior design of assembly residence building (Baoshan District, Shanghai City).

(3) Interior design of Alibaba youth apartment (Yuhang District, Hangzhou City).

(4) Interior design of SOHO youth apartment (Jiangbei New Area, Nanjing City).

(5) Architectural and environment design of "Alshan house" homestay demonstrative area (Alshan Scenic Spot, Inner Mongolia Autonomous Region).

(6) Renovation design of teaching building in South China University of Technology (Guangzhou Higher Education Mega Center).

(7) Exhibition design of China International Import Expo (Qingpu District, Shanghai City).

"室内设计6+"2019（第七届）联合毕业设计特色教育创新项目纲要

中国建筑学会室内设计分会（以下简称"室内分会"）依据《"室内设计6+"联合毕业设计特色教育创新项目章程》和相关工作要求与安排，经参加高校、命题单位等协商，形成《"室内设计6+"2019（第七届）联合毕业设计特色教育创新项目纲要》，指导开展本届联合毕业设计工作。

一、总命题：绿色营造——建筑室内装配化设计

二、项目地点：详见命题单位建筑室内装配化设计项目

三、指导单位：中国建筑学会

全国高等学校建筑学学科专业指导委员会

教育部高等学校设计学类专业教学指导委员会

四、主办单位：中国建筑学会室内设计分会

五、承办高校：北京建筑大学（开题调研）

浙江工业大学（中期检查）

南京艺术学院（毕业答辩）

六、主编单位：中国建筑学会室内设计分会

北京建筑大学

七、参加高校（学院＼专业）：

同济大学（建筑与城市规划学院＼建筑学）

华南理工大学（设计学院＼环境设计）

哈尔滨工业大学（建筑学院＼建筑学、环境设计）

西安建筑科技大学（艺术学院＼环境设计）

北京建筑大学（建筑与城市规划学院＼环境设计）

南京艺术学院（工业设计学院＼艺术与科技）

浙江工业大学（建筑工程学院＼建筑学、设计艺术学院＼环境设计）

八、命题单位：中国建设科技集团绿色建筑设计研究院

华东建筑集团股份有限公司上海现代建筑装饰环境设计研究院有限公司

杭州国美建筑设计研究院有限公司

南京观筑历史建筑文化研究院

阿尔山市文化旅游体育局

上海全筑建筑装饰集团股份有限公司

常州霍克展示系统股份有限公司

九、支持单位：东易日盛家居装饰集团股份有限公司

十、项目观察：室内分会"室内设计6+"2019（第七届）联合毕业设计活动（华北、华东、华南、华西、西北、东北地区）观察员

室内分会特邀亚洲建筑与城市设计联盟（AAUA）观察员

十一、出版单位：中国水利水电出版社

十二、时　　间：2018年10月—2019年11月

十三、活动安排：见下表

序号	阶段	时间 （年／月／日）	地点	活动内容	相关工作
1	命题研讨	2018/10/27—30	2018室内设计二十八届（重庆）年会重庆商学院	● 10月27—30日2018室内设计二十八届（重庆）年会专题展览 ● 10月27日报到；"室内设计6+"2019（第七届）联合毕业设计命题研讨会	● 室内设计（重庆）年会"室内设计6+"2018（第六届）联合毕业设计专题展览 ● 往届项目总结交流；"室内设计6+"2019（第七届）联合毕业设计总命题、子课题研讨；商讨《绿色营造卷——建筑室内装配化设计》编辑出版、年会展览等相关工作；形成本届《项目纲要》要点；确定本届活动承办高校
2	教学准备	2018/10/31—2019/02/28	参加高校、命题单位	● 联合毕业设计教学工作准备	● 参加高校报送参加本届联合毕业设计师生名单 ● 编制本届《项目纲要》；参加高校、命题单位等反馈意见与建议 ● 依据《2019框架任务书》，参加高校、命题单位联合编制、报送《2019详细任务书》 ● 毕业设计课题相关文献检索；开题调研；完成调研报告 ● 参加高校编制开题调研成果：《绿色营造卷——建筑室内装配化设计》书稿开题调研页面、调研汇报PPT ● 开题调研活动筹备
3	开题调研	2019/03/01—03	北京建筑大学	● 3月1日，报到 ● 3月2日，开题仪式、专题讲座、开题调研 ● 3月3日，项目考察	● 开题仪式；室内分会颁发本届项目导师、观察员聘书 ● 专题讲座；室内分会颁发专题讲坛演讲人聘书 ● 开题调研汇报专家点评；提交开题调研成果 ● 考察雄安新区市民服务中心、新区设计中心等项目
4	方案设计	2019/03/04—2019/04/18	参加高校、命题单位	● 方案初步设计	● 参加高校、命题单位协同组织落实毕业设计子课题相关教学环节工作 ● 明确方案初步设计目标，完成相关图纸、图表、文本编制等中期成果 ● 编制中期检查成果：《绿色营造卷——建筑室内装配化设计》书稿中期检查页面、中期检查方案设计汇报PPT ● 中期检查活动筹备
5	中期检查	2019/04/19—21	浙江工业大学	● 4月19日，报到 ● 4月20日，中期检查 ● 4月21日，项目考察	● 中期检查汇报，专家点评；提交中期检查成果 ● 考察杭州亚厦中心项目
6	深化设计	2019/04/22—2019/05/30	参加高校	● 方案深化设计	● 明确方案深化设计目标，完成相关图纸、图表、模型、文本等相应成果 ● 编制毕业答辩成果：《绿色营造卷——建筑室内装配化设计》书稿毕业答辩页面、毕业答辩汇报PPT、年会展板 ● 毕业答辩、表彰奖励活动筹备

序号	阶段	时间 （年／月／日）	地点	活动内容	相关工作
7	毕业答辩	2019/05/31— 2019/06/02	南京艺术学院	● 5月31日，报到 ● 6月1日，毕业答辩 ● 6月2日，颁发证书、活动总结、"设计再造"体验活动、项目考察	● 毕业答辩汇报，专家点评；提交毕业答辩成果 ● "室内设计6+"2019（第七届）联合毕业设计表彰奖励；活动总结 ● 室内分会"设计再造"体验活动 ● 考察南京四方当代美术馆项目，参观建筑实践展
8	编辑出版	2019/06/03— 2019/11/13	参加高校、命题单位、出版单位等	● 7月31日前，《绿色营造卷——建筑室内装配化设计》书稿内容编写、年会展板编制 ● 8月31日前，《绿色营造卷——建筑室内装配化设计》书稿统稿、排版 ● 9月20日前，《绿色营造卷——建筑室内装配化设计》审稿、加工、校对 ● 10月20日前，《绿色营造卷——建筑室内装配化设计》印刷发行 ● 10月31日前，室内分会（上海）年会专题展览、命题研讨会筹备	● 参加高校和相关专家提交《绿色营造卷——建筑室内装配化设计》书稿相应内容 ● 参加高校提交年会展板文件 ● 室内分会和北京建筑大学负责《绿色营造卷——建筑室内装配化设计》统稿排版工作 ● 出版单位负责《绿色营造卷——建筑室内装配化设计》审稿、加工、校对、印刷、发行等工作 ● 室内分会（上海）年会"室内设计6+"2019（第七届）联合毕业设计专题展览筹备 ● "室内设计6+"2020（第八届）联合毕业设计命题研讨会筹备
9	展览交流	2019/11/14—17	2019室内分会二十九届（上海）年会室内设计教育论坛	● 11月14日，报到 ● 11月15日，"室内设计6+"2020（第八届）联合毕业设计命题研讨会暨《绿色营造卷——建筑室内装配化设计》出版发行仪式 ● 11月15—17日，室内分会年会专题展览	● 室内分会（上海）年会"室内设计6+"2019（第七届）联合毕业设计专题展览 ● 《绿色营造卷——建筑室内装配化设计》发行仪式 ● "室内设计6+"2020（第八届）联合毕业设计总命题、子课题研讨 ● 亚洲室内设计联合会（AIDIA）设计教育成果交流

靳丽颖[1]、陈静勇[2]

1 北京建筑大学设计艺术研究院

2 通信作者：北京建筑大学设计艺术研究学院，教授，chenjingyong@bucea.edu.cn

艳度对比与点线面体——《绿色营造卷——建筑室内装配化设计》书籍设计

Brightness Contrast and Point-line-surface-body —— Book Design of Green Construction —— Building Indoor Assembly Design

「室内设计 6+」2019（第七届）联合毕业设计

"Interior Design 6+"2019(Seventh Year) Joint Graduation Project Event

书籍封面设计是书籍设计中的重要环节，封面是书籍的门面，与书籍的内容息息相关。当读者面对书籍的时候，首先看到的就是封面，封面设计的吸引力直接影响读者的购买意向。所以，在进行书籍封面设计构思时，就要确保封面和内容的相符，以简洁的设计语言来吸引读者的注意力。

书籍封面的元素包括色彩、图形以及文字。色彩是先于文字和图形打动读者的视觉第一语言，具有强烈的表现力。每本书的主色调应该符合文字内容的基本情调和作品的风格。

中国建筑学会室内设计分会将"绿色营造——建筑室内装配化设计"作为"室内设计 6+"2019（第七届）联合毕业设计的总命题，根据总命题的内容，可以提取"绿色"和"装配化"作为《绿色营造卷——建筑室内装配化设计》（以下简称《绿色营造卷》），书籍封面设计的图形元素，使得封面设计与本书主题紧密切合。

绿色在黄色和蓝色之间，视觉上更加柔和舒适，可以带给人安全、稳定、柔和的感觉。

任何视觉艺术，不管其形态多么复杂，变化多么丰富，归结起来，都是在体现黑、白、灰之间的相互关系。黑、白、灰可以在设计作品中体现明暗关系。本书外封选择 CMYK 色值中 K 均为 0 的三种不同艳度对比的绿色系阶梯关系，

点、线、面、体是构成设计语言的基本要素，一切空间的存在，都是由点、线、面、体组成的。点移动成线，线移动成面，面移动成体，它们互相交织、融合、衬托，形成一种秩序和韵律。书籍封面就是将色彩、文字、图形等通过点、线、面的组合与排列构成的。根据《绿色营造卷》的"装配化"主题提取非具象的"点线面体"组合图形，使人产生视觉上的区别和差异，以高低、远近、疏密、繁简、虚实的表现形式，体现装配的含义，使画面具有变化和节奏感，主次分明，形象更为突出。

色彩元素——艳度对比

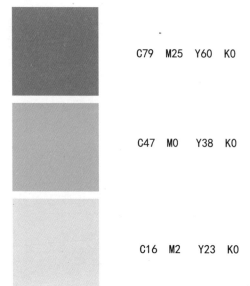

基础色

C79　M25　Y60　K0

C47　M0　Y38　K0

C16　M2　Y23　K0

图形元素——点线面体

点

线

面

体

体

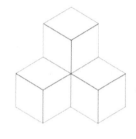
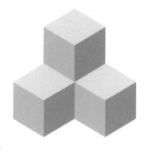

组合

封面设计

封底设计

《绿色营造卷》书籍封面设计思路为由线到面再到体的演化过程，提取六边形为主要图形元素，围绕装配化的定义，将六边形组成交织错落的立体图形，从而使书籍封面给读者以强烈的视觉冲击。同时，将三种不同艳度的绿色融入体和面中，形成不同形式的立体图形，增添多种画面效果。

本书封面的书名字体采用较宽字体，突出展现书名内容，起到吸引读者注意力的作用。封面是将六边形组合为阶梯式上升的形态，以线、面、体的表现形式为主。

封底设计是对封面、书脊的延展、补充、总结或强调，与封面紧密关联，相互映衬。封底含有条形码、书号、书籍价格等重要信息，所以在图形设计上采用简洁、对比度低的色彩设计来凸显本书主题，同时呈现由封面"体"的形态转变为封底"线"的形态的完整视觉效果。

风采定格

书脊设计

　　书脊上印有书名、册别和出版社等文字信息，方便读者查找以及归类。书籍设计采用了将封面、封底、书脊、图形连接成一个整体连续的外封形式。

　　外封设计围绕本卷"绿色"和"装配化"的主题，将"点线面体"的元素运用其中，由封面的"体"过渡为封底的"线"，同时将"点"元素运用其中，从而形成一个整体。外封整体采用对称的图形，将设计元素以书脊为左右对称轴同形同量地均匀分布，给读者以稳定、端庄的视觉感受。同时，加强设计元素的位置规则，给读者带来视觉上的平衡感，突出体现《绿色营造卷》"装配化"的主题。

　　章节隔页设计与外封设计相呼应，将"点线面体"的元素运用其中。章节隔页左页以"体"的表现为主，右页以"线"的表现为主，章节编号字体具有立体感，并且做逐章位置变化，更显生动活跃，体现装配多样化的内涵。

外封设计

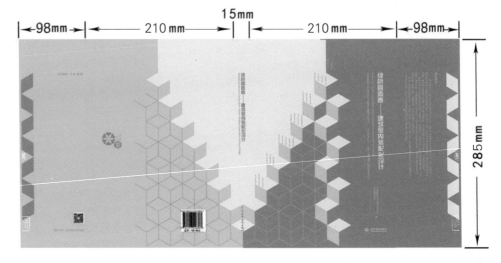

正文版式

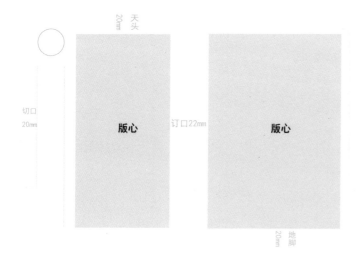

正文的版式设计以线的形式为主，增加视觉的流动性，简洁明了，与正文内容协调，不独立于正文之外，突出体现正文内容，方便阅读。

《绿色营造卷》中的图片和文字较多，因此以竖式网格布局排版为主，使书籍版面看起来更规整简洁，便于欣赏和浏览。

文字元素

基本字体

字号标准

宋体 Regular

ABCDEFGHIJKL

MNOPQRSTUVWX

YZ

abcdefghijkl

mnopqrstuvwx

yz

绿色营造
Green construction

绿色营造
Green construction

黑体

ABCDEFGHIJKL

MNOPQRSTUVWX

YZ

abcdefghijkl

mnopqrstuvwx

yz

绿色营造
Green construction

Calibri Regular

绿色营造
Green construction

拓展字体

百度综艺简体

方正清刻本悦宋简体

绿色营造
Green construction

联合指导　服务需求

热点命题　纷显特色

开题调研

「室内设计6+」2019(第七届)联合毕业设计

"Interior Design 6+"2019(Seventh Year) Joint
Graduation Project Event

问题导向

深入实地 精细考察 体验建筑与环境
收集现状资料 开展实态分析 发现设计问题

同济大学
Tongji University

小户型租赁住宅装配化设计——深圳『城中村』住宅改造
Assembly Design of The Small Rental Housing—"Urban Village" Housing Renovation in Shenzhen

『室内设计 6+』2019（第七届）联合毕业设计
"Interior Design 6+"2019(Seventh Year) Joint Graduation Project Event

小组成员

黄曼姝　王萌　詹强　周宽

基地背景分析 ┃ Base Background Analysis

城中村问题在深圳一直是个热点话题。随着深圳的快速发展，大量农村和城市郊区的土地被并入市区，外来务工人员涌进城市。但由于城市配套建设未能跟上发展速度，务工人员谋求较低生活成本，城中村应运而生，成了深圳当地的一大特色，由此也带来了一系列问题。本次设计项目的基地位于深圳白石洲的一个城中村。白石洲是深圳最大的农民房集中区，环境复杂。

白石洲鸟瞰

白石洲区域建筑密度高、容量大，作为传统房屋租赁的集中区，有大约 15 万人居住于此，存在十分庞大的租客群体，生活气息浓厚。

但随着周边科技园、华侨城、世界之窗等高端区域建设，白石洲的风貌问题愈加突出，如何以此为例进行城中村改造成了核心问题。

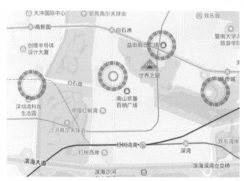

白石洲及周边

华侨城　　　　　　　　　　　　　　　　世界之窗

基地周边交通情况 ┃ Traffic Around the Base

基地周边的交通区位优势明显。在公共交通系统方面，基地周边 1 km 范围内，覆盖公交站点达 14 个，3 条地铁线路穿行而过，对周边居民而言十分便利，工作和基本生活也比较便捷。

在路政交通方面，基地北侧紧邻贯穿深圳东西的主干道深南大道，向南距离滨海立交系统也不远，且周边各层级道路布置有序，车行较为通畅。

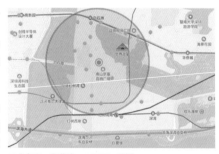

● 公交站　◎ 基地　　1km 范围

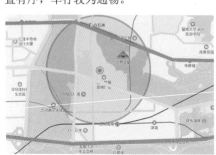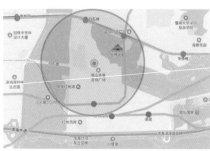

—— 主干道　次道路　支路　◎ 基地　　1km 范围　　　　—— 地铁线　● 地铁站　◎ 基地　　1km 范围

37 号楼的环境现状 | Environmental Status of Building No.37

绿色营造卷——建筑室内装配化设计

基地及周边平面

项目位于白石洲南区，东侧为街区主要道路，37 号和 25-1 号楼的首层现为沿街商铺，其余楼层及其他各楼均为民宅。

沿城中村主路向北看

建筑共 7 层，紧邻白石洲主街，从这张照片中可以看到主街商铺密集，人流量大，出入口也较为混杂。

沿城中村主路向南看

整体街道环境较为杂乱，建筑与建筑、建筑与街道之间的关系不够和谐。

存在的问题 | Existing Problem

安全性

区域环境杂乱，楼宇间距过小。

消防设施简陋，建筑结构存在隐患。

运营管理模式有待进一步的提升。

功能性

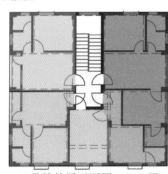

37 号楼的原平面图（2～4 层）

布局不合理　　物理环境差

功能适应性差　居家需求未满足

舒适性

色彩过于单调

材质十分简陋

未考虑人体工效

开题调研

025

客户群体分析 ∣ Customer Group Analysis

30 岁以上：16.95%
26 ～ 30 岁：50.85%
20 ～ 25 岁：32.2%

年龄组成

初中及初中以下：0%
高中：0%
博士：5.08%
大专：6.78%
本科：30.51%
硕士：57.63%

学历

做生意：0%
其他：0%
自由职业：8.47%
上班族：91.53%

职业

调研人群中 90 后占比超过 80%，高校毕业生居多，租赁需求大。

目标人群 ∣ Target Group

90 后租户

90 后租房人群已经占比首次超过 80 后，成为租房主力。
90 后对于租房最关心的问题已不再是租金，而是生活品质。
17.55% 的毕业生愿意拿出薪资的 30% ～ 50% 去租房。（《2017 年高校毕业生就业及租房趋势报告》，http://www.xinhuanet.com//local/2017-08/24/c_1121535198.htm.）

90 后租房需求和遇到的问题

要安全
要和不认识的人合租，感觉被打扰和不安全

要色彩
色彩单调，不能自己动手改造，无聊

要学琴
想学钢琴，没有地方放琴

要养猫
想养宠物，没有地方给猫猫玩

要放松
学习工作了一天好累，没有地方可以伸展筋骨

要收纳
想买的东西很多，买回来却不知道放哪里

要个性
想住"ins 风"的房子，想住波西米亚民族风的房子，想住极简风的房子……

要运动
想减肥运动，又是雾霾天又没钱去健身房，家里也没有地方运动

90 后特点分析

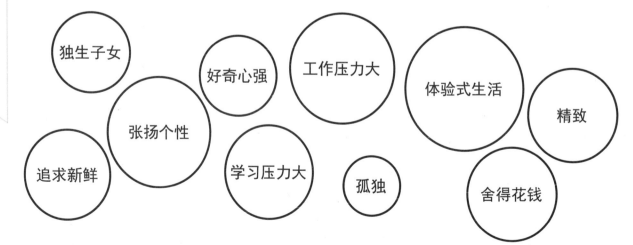

独生子女　好奇心强　工作压力大　体验式生活　精致　张扬个性　追求新鲜　学习压力大　孤独　舍得花钱

　　相比于长辈，90 后的生活水平大大提高，他们大多是独生子女，有很强的独立意识，追求新鲜、张扬个性、好奇心强。由于生活水平提高，90 后的生活品质也显得更精致，他们更加追求生活体验。但更高的学习和工作压力，以及人际交往的缺乏，也导致了这一代人孤独的现状。

「室内设计 6+」2019（第七届）联合毕业设计
"Interior Design 6+"2019(Seventh Year) Joint Graduation Project Event

90 后家庭特点

3～5岁 2.37%

5岁以上 0.63%

60% 0～3岁

37% 孕育中

90 后年轻家庭的孩子多为**低龄幼儿**

最近一周，父母与孩子交流时长？

34.9% ～2小时

30.7% 1～2小时

28% 3～5小时

23.5% 3～5小时

18.2% ～2小时

11.6% 3～5小时

80后父母

90后父母

70后父母

90 后父母愿意花更多时间**陪伴孩子**

婚育现状

家庭观

租房问题 &需求

理想家

收纳 孩子的用品太多，已经快把家里空间占满了

个性 想要属于自己风格的家，不能自己设计，不喜欢

独处 帮孩子带娃，想要个舒服的老人卧室

交流 家里视线不好，一个人带娃却不能时刻照看到孩子

多功能 想要个大大的书房，可以工作，可以睡觉，还可以看电影

独处 想要自己的卧室，可以放很多书很多玩具

活动空间 想陪孩子玩游戏，没有地方放玩具

社交 周末想请朋友小聚，家里太小，餐桌不够放

居住类型分类 ∣ Residential Type Classification

单人独居

情侣/夫妻同居

朋友分室合租

90 后上班族夫妻二人之家

母亲全职的萌娃之家

老人带娃的复合家庭

学前儿童的三口之家

设计策略 ∣ Design Strategy

安全

安全可靠 私密不被打扰

灵活

多用灵活 个性定制

共享

功能共享 社区感

「室内设计 6+」2019（第七届）联合毕业设计
"Interior Design 6+"2019(Seventh Year) Joint Graduation Project Event

同济大学
Tongji University

小户型租赁住宅装配化设计
Assembling Design of Small Rental Housing

小组成员
华立媛
佐藤辉明
肖晓溪
王安琪

人群及基地现状 | Population and Base Status

本次设计以 37 号建筑为对象。居住组团内建筑物间距较小，组团出入口较多且与沿街商业混合，作为居住空间，缺乏标示性与社区感。

功能改造： 将首层临街部分设置为商铺；
2～4 层为一层 8 户公寓；
5～7 层为一层 4 户公寓；
8 层为员工宿舍及晾晒区。

形象更新： 立面设计；楼梯间走道设计；
公共活动区设计

装配化设计： 配化模块设计；家具部品设计

光环境设计： 日光模拟分析；室内照明设计

个体
痛点

功能设计缺细分

窗口设计不合理

收纳空间不足够

晾晒空间不满足

群体
痛点

公寓楼下即商业街道
缺乏标示性和居住感

建筑形象陈旧普通
缺乏归属感和领域感

公共交往空间消极
缺乏社交性和愉悦感

客群调研与描摹 ┃ Customer Survey and Description

职 业

■白领 ■创业 ■自由职业 ■其他

其他 6%
自由职业 14%
创业 9%
白领 71%

消 费

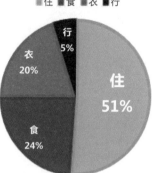

■住 ■食 ■衣 ■行

行 5%
衣 20%
食 24%
住 51%

（一家三口）独居和年轻化为租房关键词
房租位列租客日常开销首位居住状态

与朋友同住：29.0%
个人独居：21.6%
与伴侣同住：18.1%
与伴侣、小孩同住：13.8%
与宠物同住：8.2%
与亲戚同住：7.6%
其他：1.7%

（二人世界）独居和年轻化为租房关键词
房租位列租客日常开销首位居住状态

与朋友同住：29.0%
个人独居：21.6%
与伴侣同住：18.1%
与伴侣、小孩同住：13.8%
与宠物同住：8.2%
与亲戚同住：7.6%
其他：1.7%

忙碌白领

陈小A 男 24岁 研究生

互联网公司 应届管培生

月入10k 承租能力中下

职场新人 加班奋斗是常态

青年情侣

王Q 女 24岁 / 陈Y 男 26岁 本科

广告行业 既是情侣又是同行

为结婚买房攒钱 承租能力中等

窝家里看电影吃饭浪漫又省钱

精致女性

Maggie Lau 女 29岁 本科

外企 品牌经理

月入18k 承租能力中上

OL连衣裙高跟鞋是标配

一家三口

90后小夫妻 小宝2岁半

制造行业 企业中层

买房前过渡 承租能力中上

照顾小宝二人世界兼顾

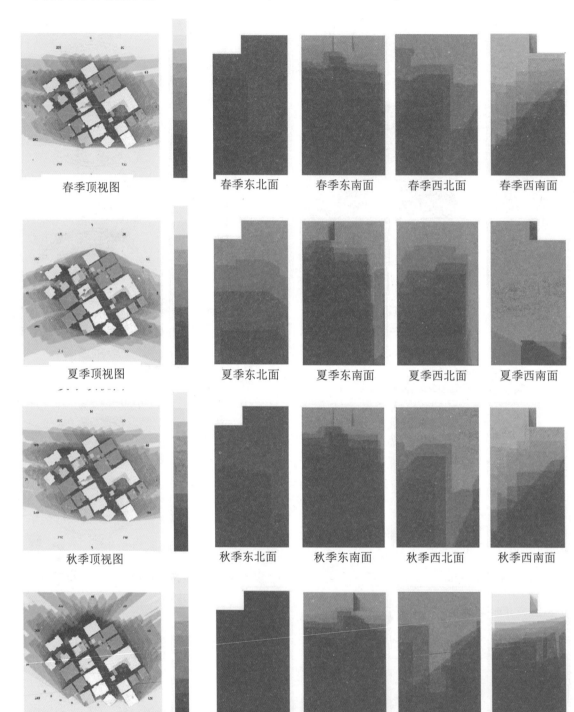

春季顶视图	春季东北面	春季东南面	春季西北面	春季西南面
夏季顶视图	夏季东北面	夏季东南面	夏季西北面	夏季西南面
秋季顶视图	秋季东北面	秋季东南面	秋季西北面	秋季西南面
冬季顶视图	冬季东北面	冬季东南面	冬季西北面	冬季西南面

基本要求
- 保证室内光环境舒适度要求
- 拓展室内功能需求

改造目标
- 环境整体形象：融入或引领?
- 对目标群体有吸引力
- 开发商的需求

灰暗

单一

混乱

压力

灰暗	群租房采光较差 老旧装修色彩暗 灰暗氛围很压抑	单一	力压成本无布置 最基本款式家具 单一色彩很无趣

混乱：室内外色彩混乱 突兀刺激易疲劳 混乱堆叠很烦躁

压力：暂时落脚群租房 无心无力去改造 一成不变很压力

明亮　　丰富　　有序　　活力

色彩体系 —— 明亮

>>> 明亮·色彩

深圳地处热带，
夏季较为炎热，
冬季较为舒适。
室内主体色应为
高反射率的冷色
调，使租客产生
凉爽清新的感觉。

白墙

白顶

浅灰砖

>>>明亮·材质

>>>硬装主体材质:
· 白色乳胶漆
· 浅色壁纸
· 浅色石塑地板

>>>软装主体材质:
· 原木
· 布艺
· 丝绒
· 织物

中和

色彩体系 —— 丰富

>>> 阳光马卡龙

>> 活力珊瑚色

2019 年度潘通流行色:
珊瑚橘　#FF6E61
明快 温暖 安心 亲密
如珊瑚礁孕育海洋生物
是情感滋养和爱的拥抱

	珊瑚	热情灿烂 →	木兰
	明黄	热烈希冀 →	朝阳
	素绿	春意盎然 →	连枝
	谧蓝	宁静安睡 →	梦美

『室内设计 6+』2019（第七届）联合毕业设计
"Interior Design 6+"2019(Seventh Year) Joint Graduation Project Event

华南理工大学
South China University of Technology

华南理工大学南校区当代艺术厅B馆改造项目
Renovation Project of Hall B of Contemporary Art Hall in The South Campus of South China University of Technology

小组成员
黄皓庭
典超华

选题意义 ┃ Significance of the Subject

当重新审视校园的公共活动空间时，发现这样一个现象：校园中存在着大量的建造完备、功能单一、利用率低、独立于大空间氛围的活动空间，它们占据校园空间的效率极低，存在缺乏管理或者是难以管理的问题。我们把它们称作"孤独空间"。

校园内的公共空间并不是仅仅考虑场地与设备，而是要更多地从空间的内涵出发，寻求发展空间。"创造联系""重塑空间"，向内部探索空间与功能的复合，去构建和营造完善的校园活动生态系统，赋予低利用率空间新的价值与内涵，这是我们倡导的途径。

调研内容 ┃ Research Contents

周边调研

发现现象

杂物堆放（临时仓库）

多个门常闭

年久失修的门

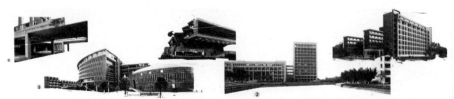

当代艺术馆B馆

确诊病情：特别的"A4 一层"

Q. A4 教学楼楼管

Q：请问为什么要把入口的门关上？
A：平时都没有人来展厅，开了门展厅的设备也不安全。
Q：为什么展厅卫生这么差，设备损坏老化，杂物堆放呢？
A：因为平时没有人来，也就不怎么管理维修。

A. 负责布展的学生

Q：今天怎么在 B 馆布展？
A：其实我们本来不想在这里展览的，原定的展厅突发漏水，所以只能选择这里。（此时正在为布展而清空 B 馆内的杂物。）

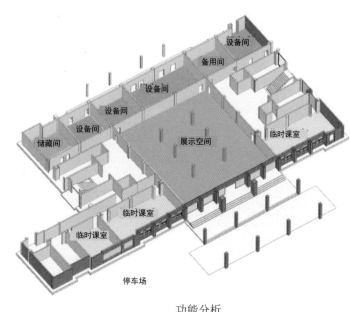

功能分析

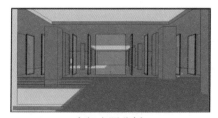

内部光照分析

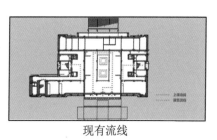

现有流线

病情诊断："自闭症"（孤立）

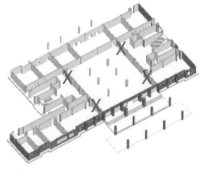

临床表现一：自我封闭（门不开）

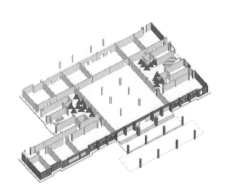

临床表现二：自暴自弃（杂物堆放）

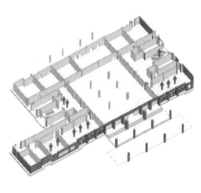

临床表现三：社交障碍
（与周围学生／功能没有内在联系）

设计定位与目标 ┃ Design Positioning and Objectives

该设计选取校园公共空间作为场地，此处存在我们所界定的"孤独空间"。在这里，我们希望改造后的建筑空间具有流通性，并且是功能复合型的。我们认为，封闭内向、远离使用群体且功能单一的建筑会造成室内空间的"孤立"。校园的各功能空间依赖物业的管理，独立且利用率不高的空间会造成额外的管理负担，同时人们为满足不同需求，需要在不同的建筑单体间来往穿梭，增加了时间和空间成本。所以，在改造设计中，在保留一定的展览空间的前提下，我们遵循原有的结构体系和一定的形式生成逻辑进行改造，根据周边环境的功能需求，将教室、教师休息室、自习休闲等功能置于其中，激活场地，营造乐活空间。

[室内设计 6+] 2019（第七届）联合毕业设计

"Interior Design 6+"2019(Seventh Year) Joint Graduation Project Event

哈尔滨工业大学

Harbin Institute of Technology

雄安设计中心室内外更新设计

Indoor and Outdoor Renewal Design of Xiong'an Design Center

小组成员

王祺叡
秦卫杰
林慧颖
李博扬

区位分析 ┃ Location Analysis

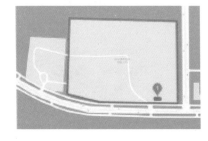

　　雄安新区位于中国河北省保定市境内，地处北京、天津、保定腹地，规划范围涵盖河北省雄县、容城、安新3个县及周边部分区域。雄安设计中心是创新设计理念的汇聚地，有中国建筑设计研究院、同济大学设计研究院、北京市建筑设计研究院、华东建筑设计研究院等国内规划设计企业入驻。

■ 雄安设计中心
■ 其他企业中心
■ 酒店
■ 员工食堂
■ 库房
■ 宿舍
■ 树林

原有建筑概念生成 ┃ Generation of Original Architectural Concept

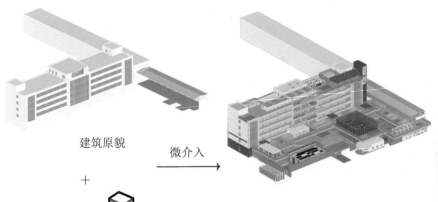

建筑原貌　　　微介入 →

+

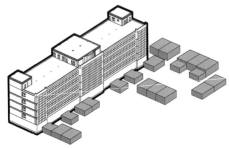

加建模块

　　尽量保持建筑原有体态，屋顶以及立面局部拆除，重点对中心部位进行更新设计。将室内外连通，加建室外平台，引入室外自然风景。使用多功能模块组建空间。

项目分层分析 |
Project Stratification Analysis

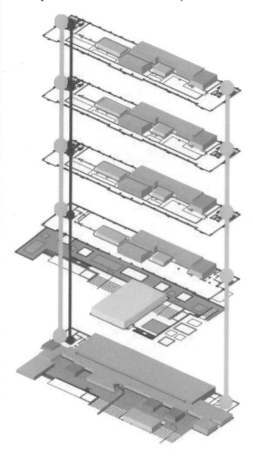

■ 主入口	■ 办公区域
■ 电梯	■ 会议室与讨论区
■ 楼梯间	■ 中央展廊
■ 绿化区域	■ 多功能模块
▨ 娱乐休闲区	●● 室内、外交通路线

室内可变模块化分析 |
Indoor Variable Modularization Analysis

专家评图区　　16 个工位　4m×4m 开放讨论区　40 人就餐区

180m² 展　示　2 个乒乓球桌　50 人报告区　　18 人封闭会议

客群分析 |
Customer Group Analysis

办公空间	25 岁	各职能齐全
休闲空间	25 ～ 35 岁	人员变动小
睡眠空间	35 岁以上	企业凝聚力
企业文化展示		

办公空间	25 岁	功能需求少
休闲空间	25 ～ 35 岁	人员变动大
睡眠空间	35 岁以上	私密要求高
企业文化展示		

光照和通风分析 |
Light and Ventilation Analysis

发现问题 |
Discovering Problem

尺度感缺失
为追求建筑整体效果，室外架空系统尺度感缺失。

新旧改造生硬
很多材料的"碰撞"都是刻意而为，不仅生硬，而且增加了成本。

缺少细分
室内空间过于开敞，视觉信息复杂，使人难以集中注意力。

开题调研

哈尔滨工业大学
Harbin Institute of Technology

阿尔山西口村民宿示范区环境设计
Environmental Design of The Demonstration Area of Villagers' Residence in Xikou Village, Alshan

「室内设计 6+」2019（第七届）联合毕业设计
"Interior Design 6+"2019(Seventh Year) Joint Graduation Project Event

小组成员
段然
张睿
谢雨萱
洪汉森

装配化现状分析 ❘ Assembly Status Analysis

现有优势

施工项目	传统设计施工阶段	装配式设计施工阶段
户型调整	主体拆改，管道改造	改加变得简单随意
铺贴墙地	瓦匠进场，工序繁多	三两工人，现场拼装
墙面油漆	油工进场，耗时烦久	干挂安装，一天完成
橱柜门窗	最便捷的一步	场内定制，材料随心
吊顶地板	工时久，废料多	基本不会产生废料
开关设置	考虑颇多，定点不易	无线开关，随意安置
洁具安装	买料安装，耗时耗力	提供 DIY 卫生间模块

在研究分析中我们发现，装配化施工有很多优势，与传统施工相比，它更加绿色环保，装配化施工现场不会出现建筑垃圾和噪声污染。作为新型施工技术，装配化施工有着良好的发展前景。通过总结，我们发现装配化施工的主要优势有：施工速度快、低碳环保、建造质量可控、建筑垃圾少、标准化机械生产。

现有问题

虽然装配化施工在我国有着美好的前景，但仍存在以下问题：

（1）技术的开发应用不如国外先进，在标准化和质量上都还有提升空间，施工技术和流程也缺乏统一标准。

（2）政府尚未出台具体激励政策，制约了大型企业的发展和建筑工业化的进步。

（3）建筑业管理体制和人才培养机制尚未适应此方面的需求，国内高等院校基本上未开辟技术工人培训渠道。

（4）目前的装配式设计未能与节能节源技术完全整合，无法契合社会的可持续发展要求。

标准体系
不完善

政府扶持
力度不够

专业人才
短缺

装配化施工的劣势

设计目标 ❘ Design Objectives

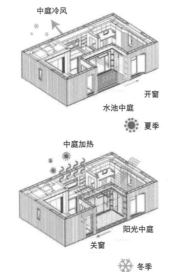

中庭冷风

开窗

水池中庭

☀ 夏季

中庭加热

阳光中庭

关窗

❄ 冬季

针对装配化的现有问题，我们提出了以下解决策略：

（1）突出设计与环境和谐共生。设计不应一味追求经济利益，设计的原则应该是"人与自然的最高和谐"。我们将设计与景观的结合作为主要的设计切入点，使设计与自然和谐共生、浑然一体。

（2）节能节源技术同绿色建筑相结合。在建设项目设计时，利用现有自然条件或植物对建筑产品进行整体布局，利用高效的自然通风、地热系统等减少空调设备的布置，建设绿色可持续建筑。

（3）简化模块设计，降低安装难度。通过简化装配模块设计，实现降低安装难度的目的，使施工人员的培训变得简单快速，从而降低人力成本。

场地现状分析 | Site Status Analysis

　　西口村东临明水河镇，西临五岔沟镇，周边旅游资源丰富，临近多个特色旅游景点和旅游城市。村周景色优美、空气清新、物产丰富，是旅游踏青的好去处。

　　西口村位于兴安盟铁路沿线和S203省道南部，省道正在扩建中，预计今年7月竣工。西口村距离机场较远，周边交通线路少，商业和旅游业的发展因此受阻。

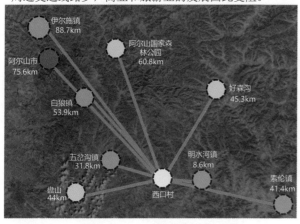

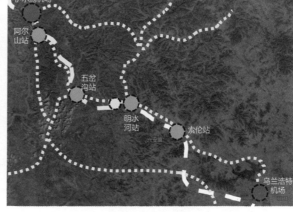

西口村至周边城镇景区示意图　　　　　　西口村周边交通道路示意图

　　西口村的道路层级简单且数量较少，有很多断头路，交通线路布局不合理。村内基础设施也十分匮乏，只有少量餐饮与小型商铺，商业种类也很单一。由此可见，西口村现有设施不能满足旅游需求，仅能满足村民的基本生活需求，无法满足游客的多元需求。

西口村城市道路与基础设施示意图

人群分析 | Crowd Analysis

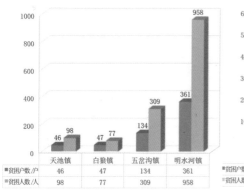

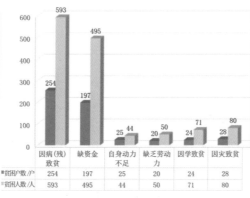

　　由左图可见，西口村地区贫困情况十分严重，且致贫原因多种多样，其中疾病和缺乏资金是主要致贫原因。为解决西口村贫困问题，政府出台了很多扶持政策以促进该地区经济发展。发展旅游业是西口村的唯一出路，建设特色民宿刻不容缓。

阿尔山市贫困人口对比分布图　　　阿尔山市致贫原因分析图

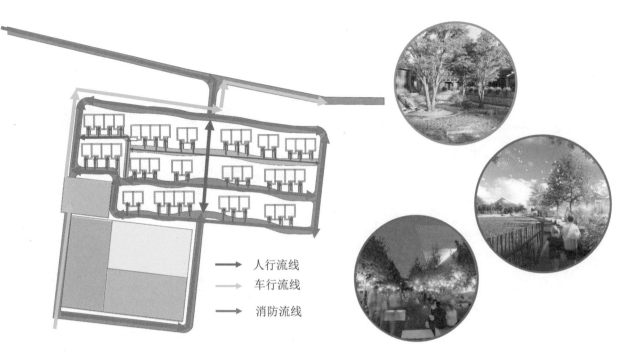

人行流线
车行流线
消防流线

场地交通流线分析图

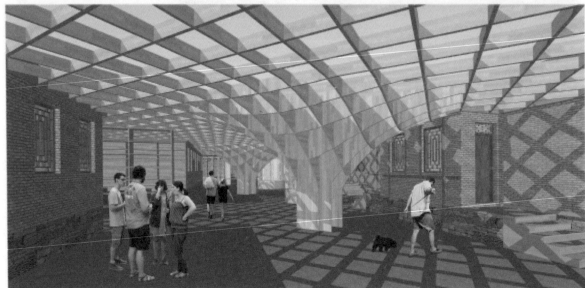

室外效果图

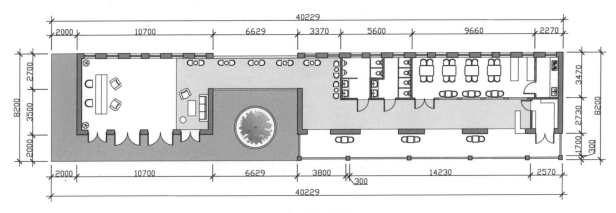

前台平面图（旺季）

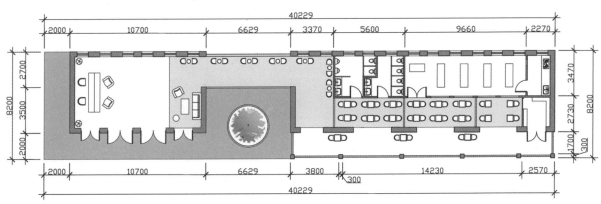

前台平面图（淡季）

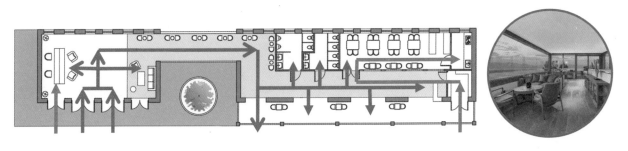

前台交通流线分析图（旺季）

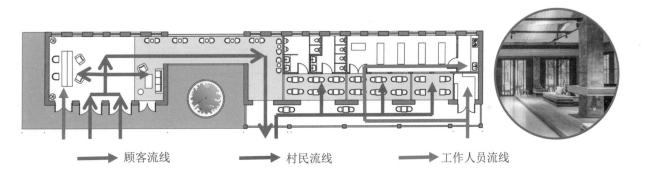

→ 顾客流线　　　→ 村民流线　　　→ 工作人员流线

前台交通流线分析图（淡季）

开题调研

西安建筑科技大学
Xi'an University of Architecture and Technology

建设可再生未来——雄安设计中心室内外环境设计
Building A Renewable Future -- Interior and Exterior Environment Design of Xiong'an Design Center

【室内设计 6+】2019（第七届）联合毕业设计
"Interior Design 6+"2019(Seventh Year) Joint Graduation Project Event

小组成员
邓莹
麦世星
秦致远
于硕

区位分析 ┃ Location Analysis

　　雄安设计中心项目位于河北省保定市容城县内，拟利用保定澳森林有限公司现有制衣大楼进行改造设计公司。厂区位于容城县澳森南大街 1 号，占地面积 158 亩。项目西至外环路内侧，东邻县政府，南邻荣乌高速，交通便捷。雄安设计中心项目位于雄安新区核心区，距离北京 100km，交通便利，区位优势明显。

场地优势 ┃ Site Advantage

　　土地规模充足，设计场地周围有较多的农田用地，位于容县西南侧附近，紧密依托容县发展的趋势。

　　改造设计场地左侧紧挨大型制衣厂，旁边分布有中海工程建设总局、中建集团等，场地与这些大型工程设计公司联系密切。

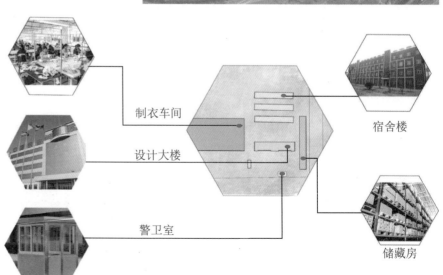

制衣车间

设计大楼

警卫室

宿舍楼

储藏房

原始大楼立面 | Original Building Facade

南立面现状

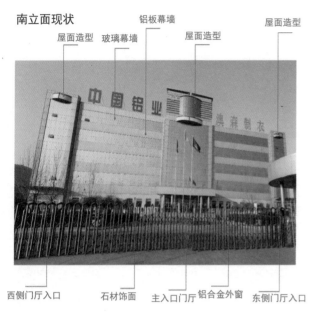

铝板幕墙
屋面造型
玻璃幕墙
屋面造型
屋面造型

西侧门厅入口　石材饰面　主入口门厅　铝合金外窗　东侧门厅入口

东西立面现状

单玻铝合金窗　小方砖饰面
小方砖饰面
单玻铝合金窗

疏散口
疏散口

周边环境分析 | Surroundings Analysis

建筑对比

老建筑办公大楼的外立面保存相对完整，结构清晰。

改造保留了大部分的外立面，少拆除，多利用。

外环境对比

原场地的景观绿化面积少，并且没有设计感。

第一次改造后的景观增加绿化。

空间分析 | Space Analysis

辅助空间　　私密空间
开敞办公空间　会议空间

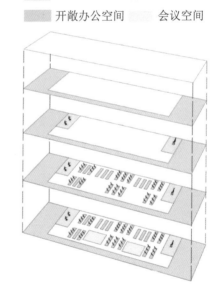

矛盾分析 | Contradiction Analysis

设计师制图工作中的矛盾：大量的设计图纸与休息睡眠需求的矛盾。

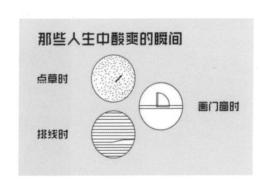

设计师灵感的缺失与急需的方案之间的矛盾。

北京建筑大学
Beijing University of Civil Engineering and Architecture

雄安设计中心零碳展示馆环境设计
Environmental Design of Zero-Carbon Exhibition Hall of Xiong'an Design Center

『室内设计 6+』2019（第七届）联合毕业设计
"Interior Design 6+"2019(Seventh Year) Joint Graduation Project Event

042

小组成员　赵佳慧　崔雨晨　柴鑫

基地概况 | Base Overview

　　本次设计为零碳展示馆，位于保定澳森制衣有限公司厂区西南角，东临澳森南大街，南临奥威路。东侧为雄安绿色建筑科学研究中心。雄安新区管委会希望将厂区改造成为设计企业聚集区。

场地调研 | Site Survey

　　左图为未改造的澳森制衣厂，从图中可看出，现零碳展示馆的位置原为一片空地，位于入口的左侧，属于交通枢纽部分。现零碳展示馆位于办公楼正前方、入口左侧，南侧有墙，无直接采光，为单层建筑，北侧为玻璃幕墙，屋顶有少量天窗。

场地分析

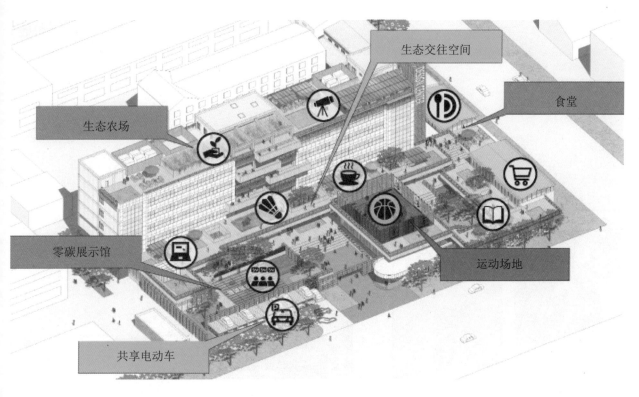

生态交往空间

生态农场

食堂

零碳展示馆

运动场地

共享电动车

能耗分析

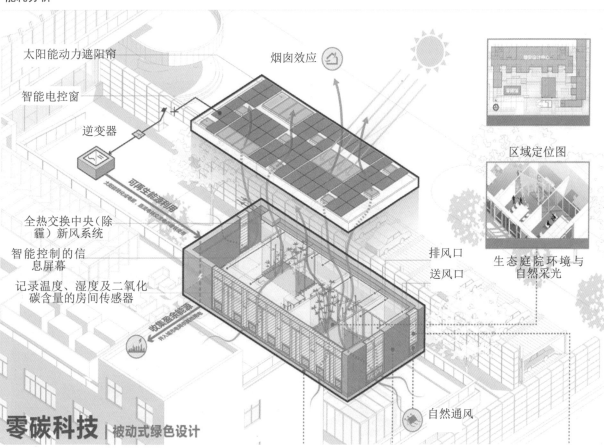

太阳能动力遮阳帘

智能电控窗

逆变器

烟囱效应

全热交换中央（除霾）新风系统

智能控制的信息屏幕

记录温度、湿度及二氧化碳含量的房间传感器

排风口

送风口

自然通风

区域定位图

生态庭院环境与自然采光

零碳科技 | 被动式绿色设计

方案设想 ┃ Conceptual Design

植根本土、回归本色的绿色设计

 绿策划　　 绿行为　　 绿能耗　　 绿循环　　 绿材料　　 绿氛围

绿色营造

窗户的设计，考虑夏季免受太阳直射，减少室内制冷所需能量；冬季将阳光引入室内空间，减少季节性的能源需求。

100% 自然通风——由电子控制的自动开合窗户系统确保内部空间全年为高质量的生活环境。

100% 自然采光——白天，每个空间都能最大限度地利用太阳光，不使用人工照明。

装配式

组合集装箱模块形成各种功能空间，单元布局与组合灵活，可生长性强，可根据发展需求进行变化。紧凑的布局也为未来发展提供了预留空间。

『室内设计 6+』2019（第七届）联合毕业设计
"Interior Design 6+"2019(Seventh Year) Joint Graduation Project Event

附件：建筑与场地图

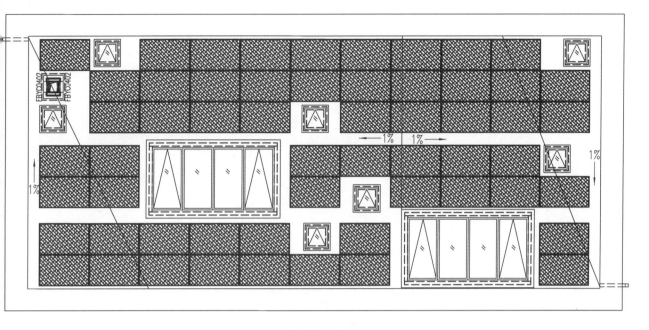

屋顶平面图

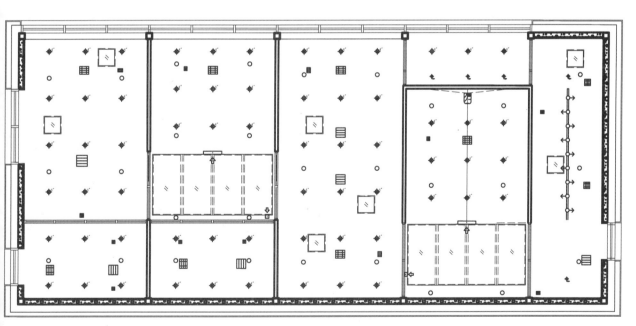

天花图

北京建筑大学
Beijing University of Civil Engineering and Architecture

内蒙古阿尔山市明水河镇西口村幸福苑民宿改造
The Renovation of Happiness Garden Homestay in Xikou Village, Mingshui River Town, Alshan City, Inner Mongoliae

「室内设计 6+」2019（第七届）联合毕业设计
"Interior Design 6+"2019(Seventh Year) Joint Graduation Project Event

046

小组成员
马宇萌
乌云塔拉
魏懋榕

基地概况 ┃ Base Overview

地理区位

"阿尔山"来源于蒙古语，有"热的圣水"之意。这里的水甘甜清冽，造就了大兴安岭夏天的郁郁苍苍、冬日的银装素裹。

阿尔山市位于内蒙古自治区兴安盟西北部，横跨大兴安岭西南山麓，是兴安盟林区的政治、经济和文化中心。它东邻呼伦贝尔市所辖扎兰屯市和兴安盟扎赉特旗，南至兴安盟科右前旗，西与蒙古国接壤，北和呼伦贝尔市新巴尔虎左旗、鄂温克自治旗毗连，总面积 7408.7km²，辖区内中蒙边境线长 93.434km。

自然资源

1. 生物资源

阿尔山市的哺乳类动物有 5 目 12 科 30 余种，禽类有 12 目 19 科 40 余种。野生植物种类繁多，共有 57 科 190 属 269 种。

2. 矿泉资源

阿尔山矿泉是世界最大的功能型矿泉之一。阿尔山周围就有冷泉、温泉、热泉、高热泉等。

3. 冰雪资源

由于特殊的地理环境，阿尔山市每年 10 月初形成有效降雪直至次年的 4 月。在长达 7 个月的冰雪期内，阿尔山冰清玉洁、银装素裹。这里雪期长、雪质好，积雪厚度平均超过 350mm，加上特殊的山形地貌，为开展冰雪运动和冰雪旅游提供了资源。

4. 森林草原

这里地处大兴安岭林区腹地，是呼伦贝尔草原、锡林郭勒草原、科尔沁草原和蒙古草原四大草原交汇处，森林覆盖率超过 64%，绿色植被率达 95%。这里地处寒温带，年均气温 -3.2℃，年降雨量 460mm，空气中负氧离子含量非常高，外国游客赞誉"空气都可以罐装出口"，是非常理想的避暑、休闲、度假、疗养的地方。

调研情况 ｜ Research Situation

阿尔山当地冰雪资源与林业资源十分丰富，调研时为冬季，虽然气温较低，但景色优美。

火墙、火炕是阿尔山地区以及北方寒冷地区的一大特色，在设计过程中考虑加入这些元素。

调研发现当地居民家中的门窗以木制为主，整体房屋较为古朴，地方特色浓郁，在设计中应考虑增加这些地方特色。

场地概况 ｜ Site Profile

设计总平面图选取二拼、三拼、四拼做设计示范，分别设计居住空间与公共空间，功能要求多样。服务设施齐全，满足各种不同人群居住需要，并考虑装配式设计手法。

阿尔山火车站建于 1937 年，是一幢东洋风格的低檐尖顶二层日式建筑，一层外壁周围是用花岗岩堆砌的乱插石墙，楼顶用赭色水泥涂盖。如今整个火车站保存完好仍在使用，是内蒙古的重点文物。在设计项目的外观上可以看出仍然带有阿尔山火车站的影子，元素可应用于室内设计。

附件：建筑与场地

北

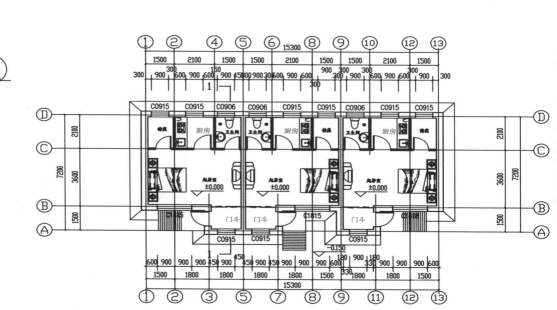

居住空间平面图（三拼）

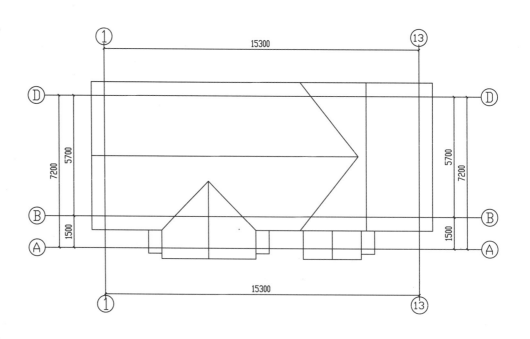

屋顶排水示意图

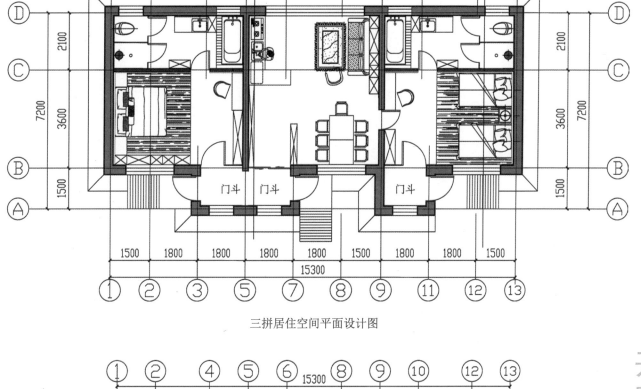

三拼居住空间平面设计图

三拼公共空间平面设计

南京艺术学院
Nanjing University of the Arts

中国国际进口博览会展示设计
Exhibition Design of China International Import Expo

「室内设计 6+」2019（第七届）联合毕业设计
"Interior Design 6+"2019(Seventh Year) Joint Graduation Project Event

小组成员
朱一丰　高榕泽
王紫荆　王紫薇
陈涛　李艺蓓
叶子萱　李赟

选题意义　｜　Significance of the Subject

　　本选题是以中国国际进口博览会的国家馆和企业馆为设计的对象，以装配化为手段的展示特装设计。中国国际进口博览会贯彻绿色理念，提出了"6R 概念"，制定了绿色展位标准，这与本届"6+"联合毕业设计的指导方向高度契合。

2018 年中国国际进口博览会概况

172 个国家、地区和国际组织参会	80 个国家馆
3600 多家企业参展	20 位国家领导人
5000 余种展品首次进入中国	130 多个国家参展
40 多万境内采购商	3600 多家企业参展
30 万 ㎡ 租馆面积	39 个交易团
80 万人次累计入场	592 个分团

场地分析　｜　Site Analysis

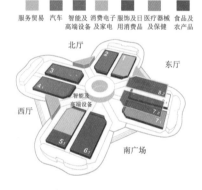

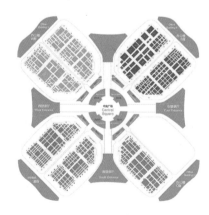

一层展厅

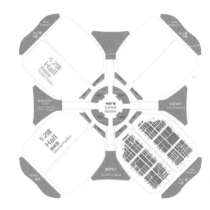

二层展厅

　　国家会展中心一层北片除 1 个大展厅为双层结构外，其余均为单层无柱展厅，单层展馆净高 32m；一层南片展厅柱网 27m×36m，净高 12m；二层的大展厅柱网 36m×54m，净高 17m。阔大的展示空间，可以让展商尽情发挥，实现高品质的形象布展。

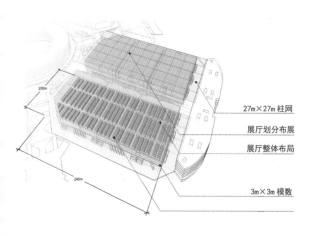

27m×27m 柱网
展厅划分布展
展厅整体布局

3m×3m 模数

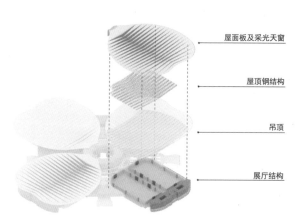

屋面板及采光天窗

屋顶钢结构

吊顶

展厅结构

承重能力 ┃ Load-Bearing Capacity

国家会展中心具有超强承重能力：一层南片的 4 个双层结构大展厅和北片 1 个小展厅的地面荷载为 3.5kN/ ㎡，二层的 5 个大展厅和 2 个小展厅的地面荷载为 1.5kN/ ㎡，一层北片的 4 个大展厅的地面荷载更是高达 5kN/ ㎡。即使是对展厅承重能力要求最高的重型机械国家会展中心亦可轻松负载。

首届进博会展馆 ┃ The First Expo Exhibition Hall

方案分析 ┃ Scheme Analysis

身处浸入式展厅的中心位置，同时也是商务洽谈的区域。

在放大镜组成的装置中穿梭，通过动作捕捉探索人体。

即是光学触碰互动的体验区，又是沉浸太空舱的最佳观看角度。

步入展厅 A，通过互动装置，了解拜耳。

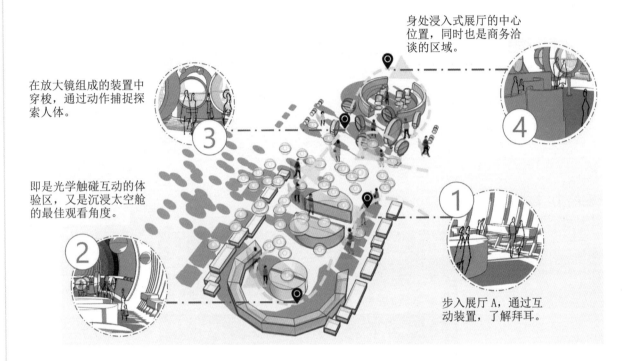

人流导向分析 ┃ Pedestrian Guidance Analysis

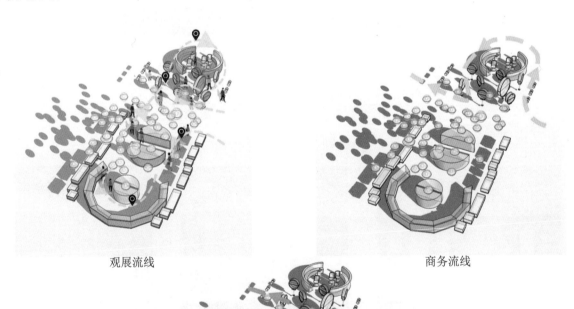

观展流线

商务流线

普通流线

场馆总平面图

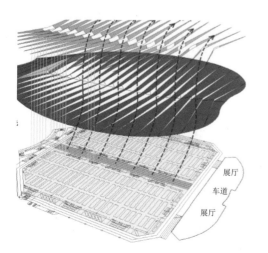

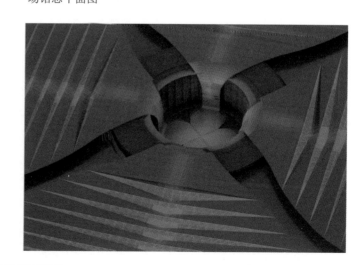

采光及排气示意图

浙江工业大学
Zhejiang University of Technology

基于一体化设计与装配化建造的阿里巴巴青年公寓室内设计研究
Interior Design Research of Alibaba Youth Apartment Based on Integrated Design and Assembly Construction

『室内设计 6+』2019（第七届）联合毕业设计
"Interior Design 6+"2019(Seventh Year) Joint Graduation Project Event

小组成员
沈令逸
张毅津

青年公寓现状分析 ┃ Current Situation Analysis of Youth Apartment

眼下青年公寓发展现状不容乐观，主要存在以下问题：
（1）人居需求提升与住居供给不足。
（2）现存青年公寓历久骤"旧"。
（3）现存青年公寓类型单一。
（4）业态变化-应对匹配性不足。
（5）对年轻人而言性，价比不高。
（6）不节能，不符合时代要求。
（7）高新科技、新型建材和结构体系应用少。
这些问题究其根本，是新型社会基础和现有建筑功能存在矛盾，新常态下的社会经济变革缺乏有效供给侧对。

产业化流程 ┃
Industrial Process

01 部件生产　工厂预制　批量生产

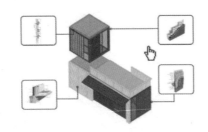

02 用户选择　自选菜单　按需组合

03 施工装配　清洁施工　高度集约

问题与策略 ┃
Problems and Strategies

（1）什么是装配化和一体化设计？
（2）如何针对阿里巴巴青年设计？
（3）适应未来的青年公寓是什么样的？

集合模块化住宅

共享空间

?

科技可变空间

服务群体分析 Ｉ Service Group Analysis

性别

女 41.90%
男 58.10%

年龄　年龄段集中在 22-29 岁

36岁以上 8.30%
30～35岁 19.80%
26～29岁 42.10%
22～25岁 29.80%

感情状况　单身率较高

5.30%　0.90%
37%　56.80%
- 单身
- 有对象，未婚
- 已婚
- 不方便透露

入职时间
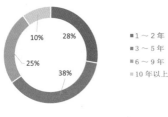
10%　28%
25%　38%
- 1～2年
- 3～5年
- 6～9年
- 10年以上

目前的居住形式　多为分室合租
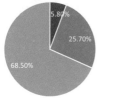
独居整套 11.10%
同室合租 5.60%
分室合租 83.30%

作息时间　加班较多，作息不规律
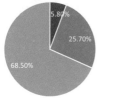
5.80%
25.70%
68.50%
- 规律
- 不规律
- 弹性作息

上班时间
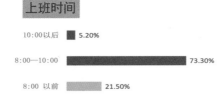
10:00以后 5.20%
8:00—10:00 73.30%
8:00 以前 21.50%

下班时间
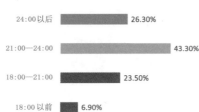
24:00以后 26.30%
21:00—24:00 43.30%
18:00—21:00 23.50%
18:00 以前 6.90%

通勤时间　通勤时间较长
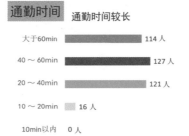
大于60min 114 人
40～60min 127 人
20～40min 121 人
10～20min 16 人
10min以内 0 人

通勤方式
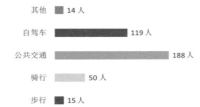
其他 14 人
自驾车 119 人
公共交通 188 人
骑行 50 人
步行 15 人

一日三餐的安排　几乎不自己做饭

外卖 46.60%
自己做饭 3%
食堂或公司周边 50.40%

有无吃夜宵的习惯
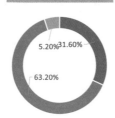
5.20%　31.60%
63.20%
- 经常
- 偶尔
- 基本不吃

休息日的安排　业余生活较丰富
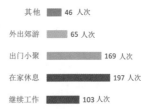
其他 46 人次
外出郊游 65 人次
出门小聚 169 人次
在家休息 197 人次
继续工作 103 人次

平时的娱乐安排　业余爱好各异
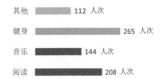
其他 112 人次
健身 265 人次
音乐 144 人次
阅读 208 人次

必需的生活空间
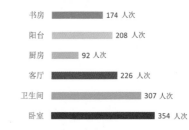
书房 174 人次
阳台 208 人次
厨房 92 人次
客厅 226 人次
卫生间 307 人次
卧室 354 人次

在目前的生活状态，会租用多大房子　多数人选择 21～60 ㎡的房子
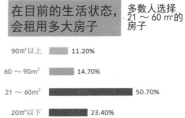
90㎡以上 11.20%
60～90m² 14.70%
21～60m² 50.70%
20㎡以下 23.40%

能否接受公共厨房等公共设施
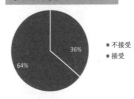
36%
64%
- 不接受
- 接受

对青年共享社区是否有居住意愿
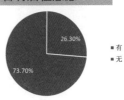
26.30%
73.70%
- 有
- 无

附件：建筑与场地图

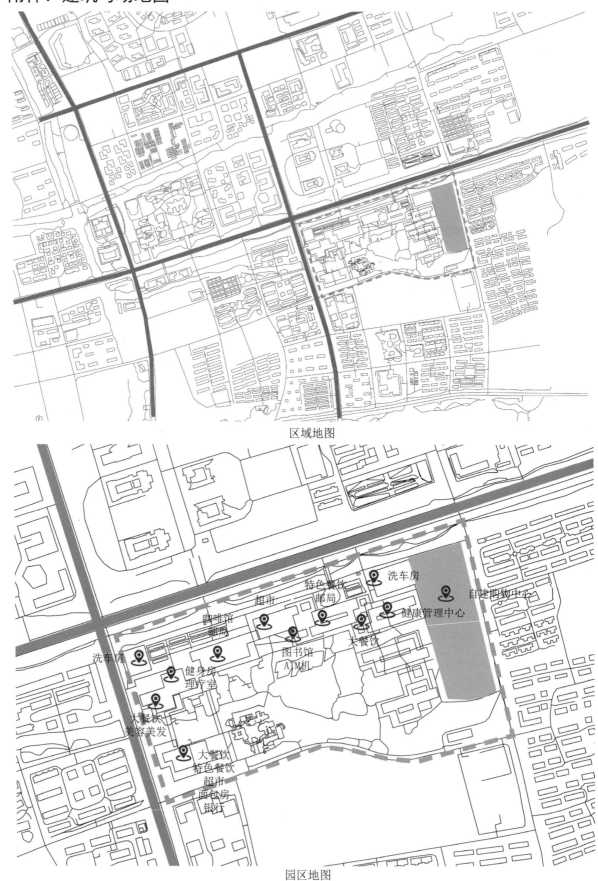

区域地图

特色餐饮
邮局
超市
洗车房
咖啡馆
药房
自建购物中心
健康管理中心
图书馆
ATM机
大餐饮
洗车房
健身房
理疗室
大餐饮
美容美发
大餐饮
特色餐饮
超市
面包房
银行

园区地图

「室内设计 6+」2019（第七届）联合毕业设计
"Interior Design 6+"2019(Seventh Year) Joint Graduation Project Event

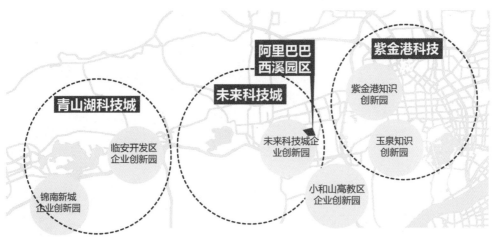

青山湖科技城

临安开发区
企业创新园

锦南新城
企业创新园

未来科技城

阿里巴巴
西溪园区

未来科技城企
业创新园

小和山高教区
企业创新园

紫金港科技

紫金港知识
创新园

玉泉知识
创新园

阿里巴巴西溪园区位于浙江省杭州市城西科创大走廊的未来科技城企业创新园中。城西科创大走廊以"创新+人才+服务"为主旨，意在打造国家自主创新示范区和世界一流、国内领先的科技创新园区。

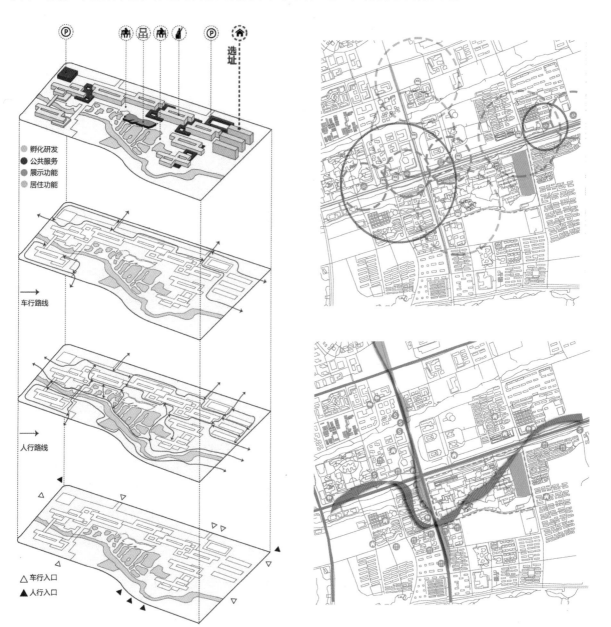

选址

孵化研发
公共服务
展示功能
居住功能

车行路线

人行路线

车行入口
人行入口

开题调研

057

浙江工业大学
Zhejiang University of Technology

基于一体设计与装配化建造的阿里巴巴青年公寓室内设计研究

Interior Design Research of Alibaba Youth Apartment Based on Integrated Design and Assembly Construction

「室内设计 6+」2019（第七届）联合毕业设计

"Interior Design 6+"2019(Seventh Year) Joint Graduation Project Event

小组成员

范诗意

陈格

选题介绍　|　Introduction of Topic Selection

本设计为"基于一体化设计与装配化建造的阿里巴巴青年公寓室内设计研究"，其关键词为"一体化设计与装配化建造""阿里巴巴青年""青年公寓室内设计"。

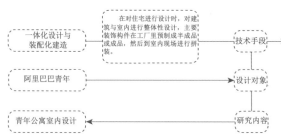

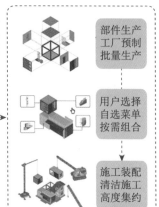

在设计过程中，我们以"一体化设计与装配化建造"为技术手段，针对阿里巴巴青年这一设计对象，进行青年公寓的室内设计研究。

青年公寓调研现状　|　Survey Status of Youth Apartments

在调研初期，我们走访了杭州 7 家青年公寓，发现市场上现有的青年公寓主要分为有共享空间和无共享空间两大类，以下是具体例子。

1. 杭州随寓青年公寓

无共享空间，产生孤独感没有交流，没有成长进步的空间。

2. 杭州 YOU+ 青年公寓

有共享空间，但共享空间与居住空间相对独立，使共享空间没有利用价值。

走访 8 家杭州的青年公寓

园区区位　|　Location of The Park

阿里巴巴西溪园区的交通可达性较强，杭州各大车站以及机场的交通便捷性较强。并且离在建的杭州火车西站仅 3km。

企业调研 I
Corporate Research

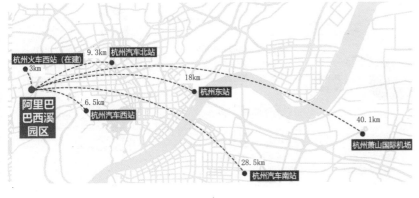

阿里巴巴西溪园区周边社区以居住功能为主。其园区东、西、南三侧均分布较多住宅小区。在园区正南方向为阿里巴巴西溪园区三期，园区北侧为浙江理工大学（现为淘宝三期规划区域）。东北方向为海创园与梦想小镇。

园区调研

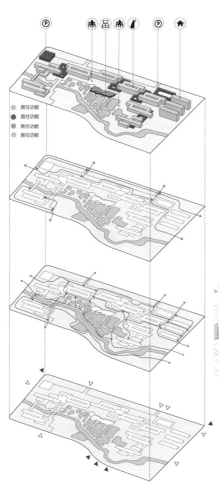

原建筑分析

冬季

夏季

室内光照   室内通风

层级一	层级一名称	层级二	层级二名称	年薪／万元
		M10	董事长（chaiman）	
P14	资深科学家	M9	副董事长（Vice chaiman）	
P13	科学家	M8	执行副总裁（EVP）	
P12	资深研究员	M7	资深副总裁（Sr.VP）	
P11	高级研究员	M6	副总裁（VP）	
P10	研究员	M5	资深总监	
P9	资深专家	M4	总监	80～100
P8	高级专家	M3	资深经理	45～80
P7	专家	M2	经理	30～50
P6	高级工程师	M1	主管	20～35
P5	中级工程师			15～25
P4	初级工程师			

企业内的职位层级主要分为P序列和M序列，其中P4～P7层级人数最多。

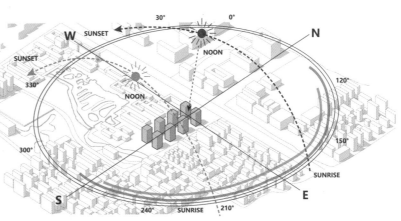

开题调研

人群分析 Ⅰ Analysis of People

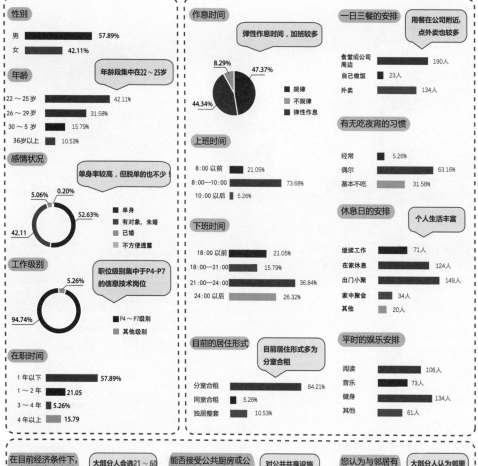

性别
- 男 57.89%
- 女 42.11%

年龄 — 年龄段集中在22～25岁
- 22～25岁 42.11%
- 26～29岁 31.58%
- 30～5岁 15.79%
- 36岁以上 10.53%

感情状况 — 单身率较高，但脱单的也不少
- 5.06% 0.20%
- 42.11 52.63%
- 单身
- 有对象，未婚
- 已婚
- 不方便透露

工作级别 — 职位级别集中于P4-P7的信息技术岗位
- 94.74%
- 5.26%
- P4～P7级别
- 其他级别

在职时间
- 1年以下 57.89%
- 1～2年 21.05
- 3～4年 5.26%
- 4年以上 15.79

作息时间 — 弹性作息时间，加班较多
- 8.29%
- 47.37%
- 44.34%
- 规律
- 不规律
- 弹性作息

上班时间
- 8:00以前 21.05%
- 8:00～10:00 73.68%
- 10:00以后 5.26%

下班时间
- 18:00以前 21.05%
- 18:00～21:00 15.79%
- 21:00～24:00 36.84%
- 24:00以后 26.32%

目前的居住形式 — 目前居住形式多为分室合租
- 分室合租 84.21%
- 同室合租 5.26%
- 独居整套 10.53%

一日三餐的安排 — 用餐在公司附近，点外卖也较多
- 食堂或公司周边 190人
- 自己做饭 23人
- 外卖 134人

有无吃夜宵的习惯
- 经常 5.26%
- 偶尔 63.16%
- 基本不吃 31.58%

休息日的安排 — 个人生活丰富
- 继续工作 71人
- 在家休息 124人
- 出门小聚 149人
- 家中聚会 34人
- 其他 20人

平时的娱乐安排
- 阅读 106人
- 音乐 73人
- 健身 134人
- 其他 61人

在目前经济条件下，会租用多大房子 — 大部分人会选21～60m²的生活空间
- 20m²以下 21.05%
- 20～60m² 50.74%
- 60～90m² 5.26%
- 90m²以上 22.95%

您必须需要的生活空间？
- 卧室 179人
- 卫生间 172人
- 客厅 83人
- 厨房 105人
- 阳台 159人
- 书房 89人

能否接受公共厨房或公共洗衣机等公共设施？ — 对公共共享设施接受程度高
- 不能接受 36.84%
- 能接受 63.16%

您对青年公寓、青年共享社区这类青年人聚居的社区是否有居住意愿？ — 愿意入住青年共享社区人数较多
- 26.32%
- 73.68%
- 有
- 无

对公共共享设施接受程度高 — 您认为与邻居有交往的必要吗？ — 大部分人认为邻里交流有必要
- 有 73.68%
- 无 26.32%

未来什么样的住宅会更加吸引您 — 科技化住宅吸引力较强
- 共享住宅 21.05
- 科技化住宅 63.16%
- 就现在这样 5.26%
- 其他 10.53%

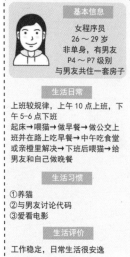

对于人群的调研一共进行了两轮。初步调研时，对目标人群的共性生活方面进行了解及数据统计。在二次深入调研中，我们对阿里巴巴的青年员工一些个性生活及个性需求方面进行了问卷调研。

此外，我们还进行了一对一的访谈以及追踪采访了某几个员工的一周生活并做了整理记录。

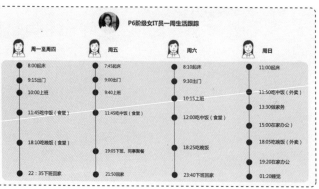

P6阶级女IT员一周生活跟踪

周一至周四
- 8:00起床
- 9:15出门
- 10:00上班
- 11:45吃中饭（食堂）
- 18:10吃晚饭（食堂）
- 22:35下班回家

周五
- 7:45起床
- 9:00出门
- 9:40上班
- 11:45吃中饭（食堂）
- 19:05下班，同事聚餐
- 21:50回家

周六
- 8:10起床
- 9:30出门
- 10:15上班
- 12:00吃中饭（食堂）
- 18:25吃晚饭
- 23:40下班回家

周日
- 11:00起床
- 11:50吃中饭（外卖）
- 13:30做家务
- 15:00在家办公
- 18:05吃晚饭（外卖）
- 19:20在家办公
- 01:20睡觉

『室内设计 6+』2019（第七届）联合毕业设计
"Interior Design 6+"2019(Seventh Year) Joint Graduation Project Event

附件：建筑图纸

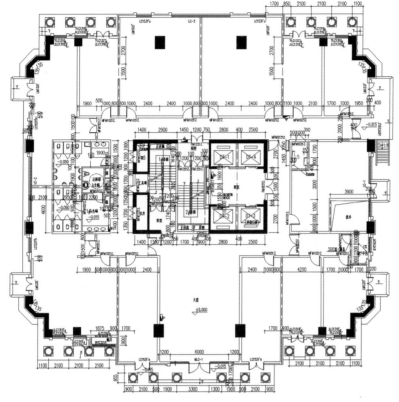

住宅首层平面图

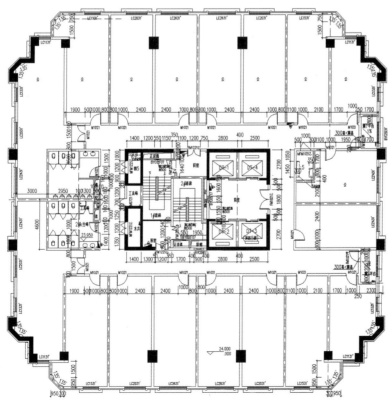

住宅标准层平面图

热点命题　纷显特色

联合指导　服务需求

中期检查

「室内设计6+」2019（第七届）联合毕业设计

"Interior Design 6+"2019(Seventh Year) Joint
Graduation Project Event

循序渐进

开阔学科视野 广泛交叉融合 遵循设计程序
讲求设计方法 探寻解决问题途径 多元启发
交流

同济大学
Tongji University

小户型租赁住宅装配化设计——深圳『城中村』住宅改造
Assembly Design of The Small Rental Housing--"Urban Village" Housing Renovation in Shenzhen

『室内设计 6+』2019（第七届）联合毕业设计
"Interior Design 6+"2019(Seventh Year) Joint Graduation Project Event

小组成员
黄曼姝
王萌
詹强
周宽

建筑立面设计 ┃ Building Facade Design

东立面：固定式垂直遮阳

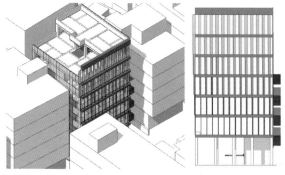

根据辐射量的大小确定格栅和立面的夹角，精确控制每个窗口的遮阳量。

西立面：设置凸窗改善采光

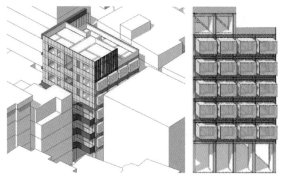

设置凸窗，增大天空的可视率，改善室内的光照环境。

公共空间设计 ┃ Public Space Design

底层公区

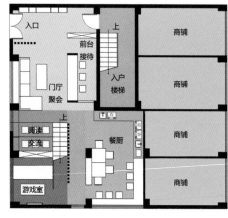

门厅聚会　阅读交流　游戏娱乐
眺望静思　交互餐厨　茶憩会晤

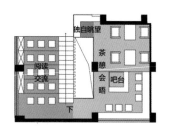

屋顶公区

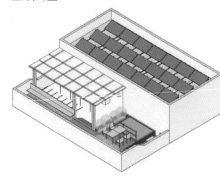

· 空间复合（晾衣杆与健身杆相组合）
· 引入水景（景观、消防、保温隔热）

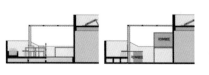

户内空间设计 | Interior Space Design

改造前后户型对比

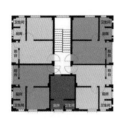

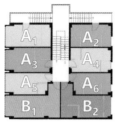

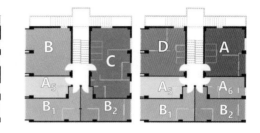

改造前标准层：
户型单一
厨卫内套，缺晾晒

改造后 2～4 层：
户型小（独居、情侣 / 分室合租）
租金少、独卫 + 晾晒

改造后 5～6 层：
户型多样（二人、三口之家、复合家庭）
品质高、独卫 + 晾晒

小户型设计

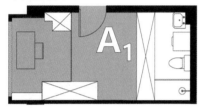

单人独居　使用面积：14.8 ㎡
　　　　　活动面积：9.5 ㎡
　　　　　占比 64%

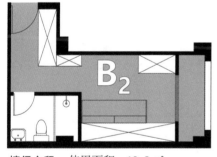

情侣合租　使用面积：19.8 ㎡
　　　　　活动面积：12.3 ㎡
　　　　　占比 62%

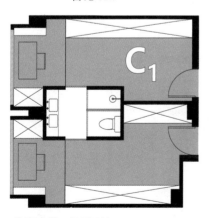

分室合租　使用面积：29.9 ㎡
　　　　　活动面积：22.7 ㎡
　　　　　占比 >76%

家庭型设计

90 后上班族夫妻二人之家

可分可合式大卧室 & 书房
户内使用面积：29.7 ㎡

母亲全职的萌娃之家

中心式活动空间
户内使用面积：31.2 ㎡

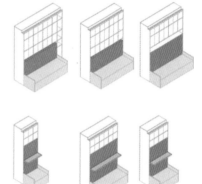

翻折床 & 沙发的收纳组合

可变卧室模块

家具模块的微更新

同济大学
Tongji University

小户型租赁住宅装配化设计
Assembling Design of Small Rental Housing

「室内设计 6+」2019（第七届）联合毕业设计
"Interior Design 6+"2019(Seventh Year) Joint Graduation Project Event

小组成员
华立媛
佐藤辉明
肖晓溪
王安琪

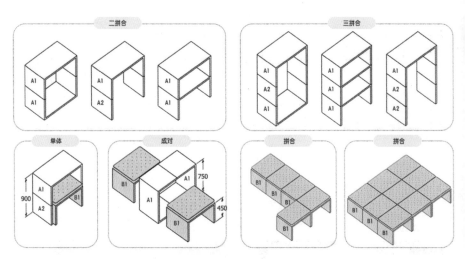

户型模块分析图

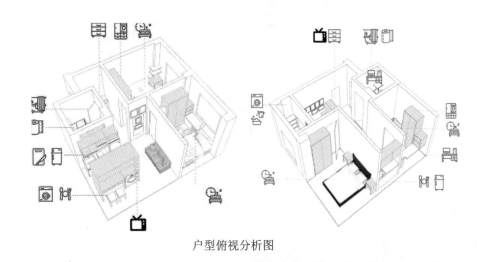

户型俯视分析图

户型平面分析图

立面设计 | Facade Design

色彩的重构设计

层次的重构设计

尺度的重构设计

立面的绿色生态化设计

中期检查

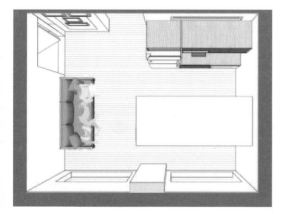

休息区使用分析图

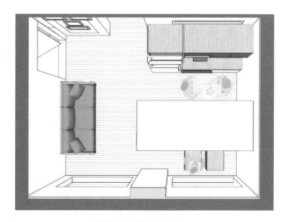

娱乐办公区分析图

067

华南理工大学
South China University of Technology

华南理工大学南校区当代艺术厅 B 馆改造项目
Renovation Project of Hall B of Contemporary Art Hall in The South Campus of South China University of Technology

【室内设计 6+】2019（第七届）联合毕业设计
"Interior Design 6+"2019(Seventh Year) Joint Graduation Project Event

小组成员　黄皓庭　典超华

概念提出 | Proposed Concept

校园"孤独"公共活动空间的改变与再塑造

　　利用"空间工具"并通过使用对象的简单操作打破空间的孤立，创造该空间与周围的联系，使空间具有"流动性"。

　　"空间工具"对空间的状况进行调节和规划，使空间在不同的环境与需求下呈现出不同的状态。

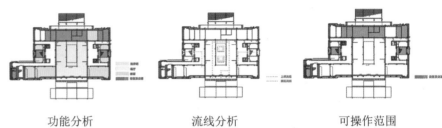

功能分析　　　　　　流线分析　　　　　　可操作范围

设计策略 | Design Strategy

第一步：
打破边界，
还原空间流动性。

第二步：
置入盒子规范空间秩序，
创造可以利用的墙面 ，
利用天井创造垂直联系。

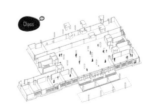

第三步：
功能复合。

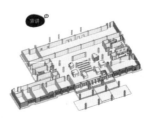
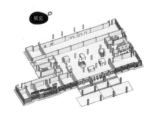

设计分析 | Design Analysis

展览模式　　　　　　　　　　工作坊模式　　　　　　　　　　宣讲模式

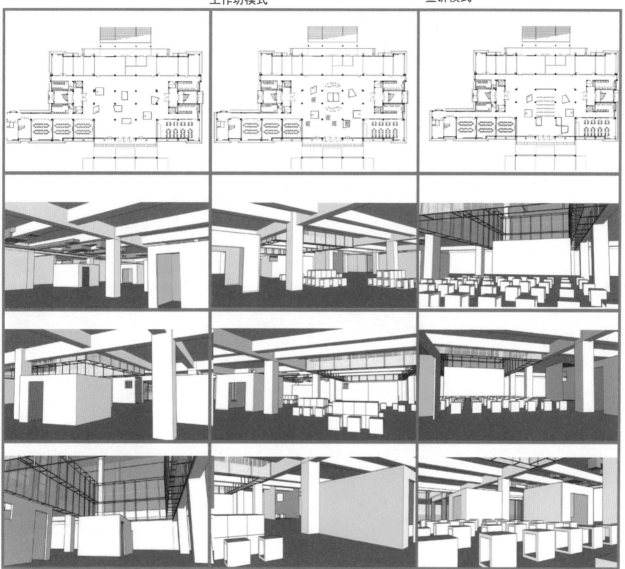

技术解析 | Resolution of Technology

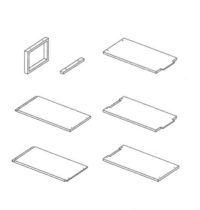

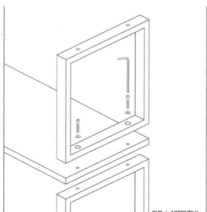

安装 | ASSEMBLY

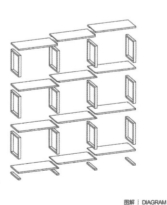

图解 | DIAGRAM

069

问题发现 | Finded Problem

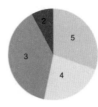

工作时长 　　　　　压力指数 　　　　　室内风格

　　我们通过网络问卷调查收集到了来自全国设计院的30位设计师的问卷。我们发现大部分的设计师是在26～35岁之间，同时发现他们的工作时长均在10～12h，压力指数在中等偏上，他们对于室内的风格更喜欢清新自然风，更期望多种的休闲方式。

　　从实际场地可以看出，棋盘式的道路分割了各种业态，缺乏多样性，整洁宽敞的街道旁很少有市民自发的活动发生，城市就少了很多市井气息。因此我们提出三个比较理想化的愿景：建立城市的归属感，塑造空间的故事性，激发人们的参与感，大家共同为雄安增添属于自己的记忆。

概念提出 | Proposed Concept

休闲模式单一
开放式办公私密性差
功能和空间布置缺乏创意
室内外联系差

重构功能空间
重构工作模式
重构人与空间的关系
重构室内与室外的联系

初步设想：一层空间——租赁

一层原空间布置　　　　　置入展览空间

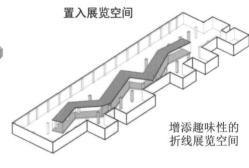

增添趣味性的折线展览空间

置入各种模块　　　　　整体空间

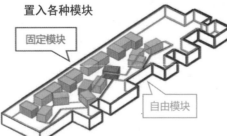
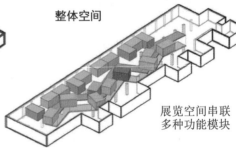

固定模块　　自由模块

展览空间串联多种功能模块

多功能模块尺寸

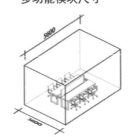
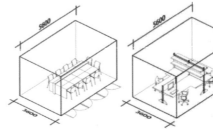

哈尔滨工业大学
Harbin Institute of Technology
雄安设计中心室内外更新设计
Indoor and Outdoor Renewal Design of Xiong'an Design Center
「室内设计 6+」2019（第七届）联合毕业设计
"Interior Design 6+"2019(Seventh Year) Joint Graduation Project Event

小组成员
王祺叡
秦卫杰
林慧颖
李博扬

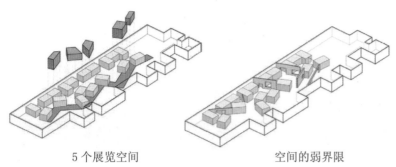

5 个展览空间　　　　　空间的弱界限　　　　　固定办公入口

附加空间　COF　墙面材料　　　　　　　　打开的模块　打破柱网

初步设想：二层空间——常驻

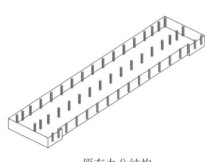

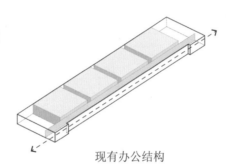

原有办公结构　　　　　　　　　　现有办公结构

　　二层作为常驻办公空间，设定为可容纳 100 人左右的设计院，暂定为 4 个所。原雄安设计中心的办公空间较为传统，室内外空间缺乏交流。

　　改造后的办公空间将中心区域留出，将办公空间分布在两侧，使室内外活动更好地融合。

　　空间上以中间柱为轴线，向两侧各退后了 4m，形成了一个 4m 宽的交通空间。同时又设置了两个功能盒子可以让同侧的两个所进行更好的交流。

功能盒子布置

视线分析

　　办公空间、专注空间和过渡休闲空间交替分布，不仅可以提高工作效率，还可以有效地释放压力，作为一个充电补给的地方。

　　视觉上，确保每个角落都可以与室外进行视线上的交流，这样也可以减缓一部分压力。

灵活可变的设计

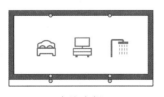

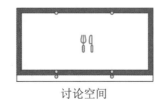

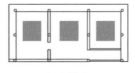

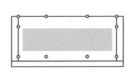

专注空间　　　　　讨论空间　　　　　会议模式　　　　　自由闲聊

哈尔滨工业大学
Harbin Institute of Technology

阿尔山西口村民宿示范区环境设计
Environmental Design of The Demonstration Area of Villagers' Residence in Xikou Village, Alshan

『室内设计 6+』2019（第七届）联合毕业设计
"Interior Design 6+"2019(Seventh Year) Joint Graduation Project Event

小组成员　段然　张睿　谢雨萱　洪汉森

概念提出 | Proposed Concept

　　我们总结了阿尔山市西口村的地域特色，发现主要有以下三点：西口村位于农林牧三界交汇处，自然景观丰富，植被富有层次；当地物种资源丰富，有牛、羊、鹿等养殖动物，还有雪兔、柳根等珍稀野生动物；西口有蒙古族、满族、汉族等多个民族，具有多元的民族风情特色。可见，西口村是一个多元文化交汇融合的地方，我们结合西口村的这个特点形成了我们的设计方案。

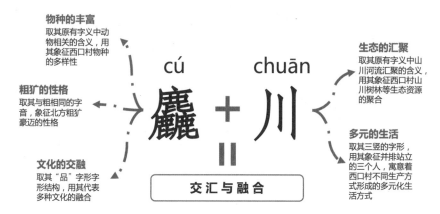

物种的丰富
取其原有字义中动物相关的含义，用其象征西口村物种的多样性

粗犷的性格
取其与粗相同的字音，象征北方粗犷豪迈的性格

文化的交融
取其"品"字字形字形结构，用其代表多种文化的融合

cú　chuān

麤 + 川 = 交汇与融合

生态的汇聚
取其原有字义中山川河流汇聚的含义，用其象征西口村山川树林等生态资源的聚合

多元的生活
取其三竖的字形，用其象征并排站立的三个人，寓意着西口村不同生产方式形成的多元化生活方式

　　我们提取麤川二字的字音、字形、字义，用其表达西口村物种丰富、性格粗犷、文化交融、生态汇聚、生活多元的地域特色，将两个字结合在一起，最终点出我们的设计主题——交汇与融合。麤川不是一个名词，而是一个动词，它表达的是一个融合的过程。我们意求通过设计的手段将当地丰富的资源和文化汇聚在一起，从而达到吸引村民和外地游客在此聚集的设计目的。

设计目标 | Design Target

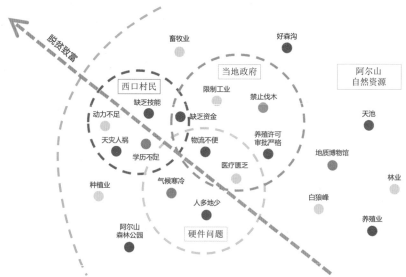

　　在设计中，我们将依靠阿尔山地区优质的自然资源，发展绿色环保的旅游业，并通过旅游业的发展突破西口村种种贫穷困难，打造一条脱贫致富之路。

业态规划 | Industrial State Planning

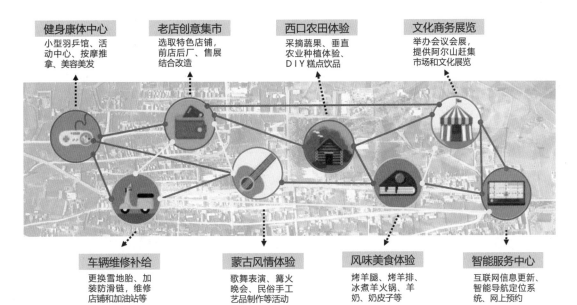

健身康体中心
小型羽乒馆、活动中心、按摩推拿、美容美发

老店创意集市
选取特色店铺，前店后厂、售展结合改造

西口农田体验
采摘蔬果、垂直农业种植体验、DIY糕点饮品

文化商务展览
举办会议会展，提供阿尔山赶集市场和文化展览

车辆维修补给
更换雪地胎、加装防滑链，维修店铺和加油站等

蒙古风情体验
歌舞表演、篝火晚会、民俗手工艺品制作等活动

风味美食体验
烤羊腿、烤羊排、冰煮羊火锅、羊奶、奶皮子等

智能服务中心
互联网信息更新、智能导航定位系统、网上预约

设计方案 | Designing Scheme

场地内置入了多个下沉小剧场，供游客和村民夏季观影、集会使用。

道路两侧的蒙古包元素艺术装置增强了方案的地域特色。

基地边设置的模块化有机农场促进了当地农产品的自产自销。

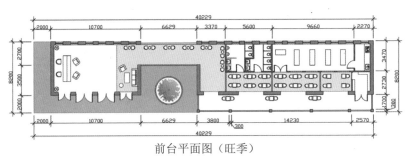

前台平面图（旺季）

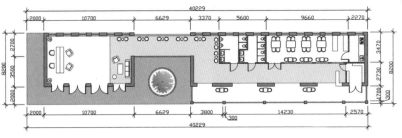

前台平面图（淡季）

双拼室内平面图

错层室内平面图

绿色营造卷——建筑室内装配化设计

中期检查

073

西安建筑科技大学
Xi'an University of Architecture and Technology

建设可再生未来——雄安设计中心室内外环境设计

Building A Renewable Future — Interior and Exterior Environment Design of Xiong'an Design Center

「室内设计 6+」2019（第七届）联合毕业设计

"Interior Design 6+"2019(Seventh Year) Joint Graduation Project Event

小组成员
邓莹
麦世星
秦致远
于硕

场地解析 | Site Analysis

  雄安设计中心总用地 6909 ㎡，建筑面积 12393 ㎡，主楼为五层，面积 11507 ㎡，预计有 12 家不同规模的设计院、公司入驻，主要承载的是设计类、工程类企业。总设计人员人数约 600 人，其中设计类工作者占四分之三，是整栋大楼人员的主要构成。

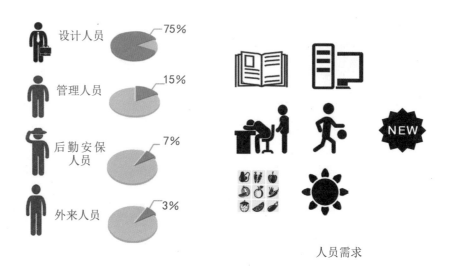

  设计人员   75%
  管理人员   15%
  后勤安保人员   7%
  外来人员   3%

人员需求

  场地功能与使用人群的转变，使场地本身与人群使用需求产生了一系列的矛盾：现有室内空间不足以承载许多休闲活动，办公桌椅不适合现在的办公需求，一些特殊的绘图活动不适合在现有办公桌椅上展开，设计师午睡以及加班之后的短暂休息没有地方满足这一需求，高强度的机械劳动导致的想法灵感缺失等。

初步构想 | Preliminary Conception

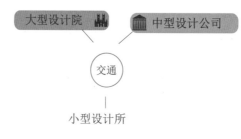

大型设计院   中型设计公司
交通
小型设计所

  雄安设计中心在改造之后我们发现，空间划分有些单一，基本每一层都被一分为二，供两个公司使用，在很大程度上限制了雄安设计中心的规模，在设计愿景上，我们希望雄安设计中心并不是只面对单一的大型设计院，而是各种规模的设计类企业都能在雄安设计中心入驻，从而扩大设计中心的使用范围，塑造一个适用性较强的设计综合体。

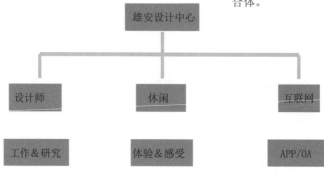

雄安设计中心

设计师   休闲   互联网

工作&研究   体验&感受   APP/OA

　　设计中心的人员主体是设计人员，年龄在 25 ～ 40 岁，男女比例大致为 3:1。在年龄结构上就决定了他们的需求是丰富多样的，会比较在乎一些丰富的休闲活动，对办公环境的需求也会相对较高，但往往办公地点难以满足这些需求，员工的办公效率以及办公的心态也会受到很大的影响，所以我们对雄安设计中心的室内外改造方向主要是面向设计师这一主要群体，提高人群关怀，增设活动空间，优化办公结构，从而让使用者能在设计中心舒适地办公。

问题解析 ┃ Problem Analysis

　　雄安设计中心落成后，经过了实地调研和分析，在设计中还是存在一些问题，虽然设计中心运用了很多先进的建筑改造理念，例如微介入。多种绿色改造手法，但由于造价等条件限制，最后呈现的效果并没有达到预期。

　　雄安设计中心在改造后，问题来源很多。空间上，室内划分较为单调，面向群体规模较为单一，空间分布与功能分配缺乏创意。在对人群的关怀上，也显得较为缺乏，其中能容纳的活动只有办公，以及一些简单的活动，活动形式较为单调，无法调动员工的工作积极性。室外环境上，能容纳的活动也较为有限，最明显的功能只有交通功能，屋顶部分并没有得到很好的利用，只做了部分卵石铺地，利用率较低。

　　因为针对的设计师这一群体的特殊性，日常工作中往往会出现很多不舒适，例如灵感的缺失，日常休息的问题，在工作展开时，基本无法进行短暂的集群交流工作，不利于一些具体工作的展开。具体部门的不同，工作类型也具有很多特殊性，往往单一的家具无法满足各种特殊的工作需求，更多的储物空间、更大的绘图空间成为了他们的现实诉求。在环境上，设计师这一群体还提出了更高的要求，也就是周围办公环境的优化，他们普遍厌恶空调，对通风、气温方面要求较高，在光线上，单一的人工光源会让他们感觉难受。

　　在设计手法上，提出了很多设计理念，但并没有得到很好的实施，室内以交通空间为界限，和室外产生了明显的分割，导致了一部分通风条件的减弱。在减少碳排放上，主要源头是能源收集与排放减少，在能源收集上，采用了很多手法，很多地方会铺设光伏发电板，但其他能源的利用上会有一些不足。部分绿植种植不太合理，部分植物接受不到光源，已经有了很明显的枯死，例如外加的廊道下种植了部分竹，但能接收到自然光的时间只有下午四五点之后，没办法得到很好的生长。

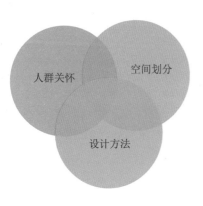

解决策略 ┃ Resolving Strategy

　　空间上进行优化和重构，以适应后期不同规模公司的入驻，营造不同空间形式，以适应不同的活动需求。
　　人群关怀上，立足于设计师群体，根据设计师日常工作中会感到不舒服的问题，通过改进办公家具、办公环境，并丰富设计师的活动形式，来提高设计师的工作效率。
　　设计方法上，主要运用装配式与绿色营造的方法，通过一系列模块的设计、组合、安装，来满足各种使用需求，以增加能源收集和减少碳排放两方面，营造一个绿色可持续的室内外环境。

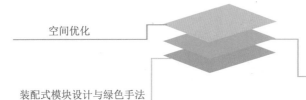

概念提出 ┃ Proposed Concept

装配式

四壁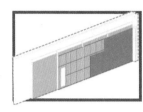

天花

地面

展陈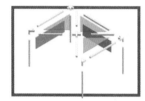

展陈设计

篇章	展览名称	展览形式	展览内容
序	传播零碳生活新概念	主视觉 logo 墙	零碳建筑总论
第一篇章	零碳技术与材料	展板、视频	展板：风帽、溶液除湿、太阳能热水器等、蓄热体等 视频（2 个）：①雨水回收、屋顶绿化；②毛细管辐射
第二篇章	零碳建筑与技术拆分	展柜（微缩模型）	零碳展览馆微缩模型（阳光房、保温墙体、中空 low-e 玻璃、粉煤灰混凝土结构、零碳地板、屋顶绿化、光伏发电、太阳能热水器、雨水回收、能耗监测系统、生物质能源）、毛细管辐射微缩系统微缩模型，给水系统微缩分析
第三篇章	零碳餐厅体验	实物（放映厅、餐厅）	在零碳餐厅里，有啤酒瓶做的灯具、冰做的酒杯、轮胎做的桌子、水管做的椅子等。食物材料均来自雄安新区的有机农场，并通过零排放汽车运到场馆，通过小功率炉灶烹制，再装入用饼干、桃子等食品制造的盘子里。用完正餐，可以把餐具当饭后甜点吃掉。没被吃掉的餐具将会和生物垃圾一起，被收集到生物能炉内，用于发电、发热。零碳馆所需的电能和热能，可以通过"生物能热电联产系统"对餐厅内各种有机废弃物、一次性餐具等降解而获得。降解完成后，最终余下的"产品"，还能用作生物肥，真正实现变废为宝
第四篇章	零碳技术虚拟体验	VR 体验	以游戏的方式来感受水源热泵等系统。椅子形态各异，材料也是五花八门：报纸、铁管、塑料桶、硬纸盒等，全是废旧回收材料
第五篇章	让零碳走进生活	体感互动	参观者不仅可以亲眼看到会发光的墙、会发电的窗户这些低碳技术，还可以通过碳测系统了解自己的节能减排记录。碳测系统是一个基于个人二氧化碳排放的网上注册系统，每一个登陆的人都可以选择他们减排的方式
第六篇章	零碳雨水花园	实物（雨水花园循环系统）	室内绿植、屋顶花园与室外绿地联在一起，由室外的蓄水池统一给水。主要针对屋面汇集雨水和地面径流雨水的处理，通过屋顶绿化、雨水花园、透水性地面、生态滞留沟、贮存回用等技术，实现对雨水的收集和利用
尾声	未来城市与生活	垂直绿化墙	零碳城市、零碳建筑、零碳家庭

光环境分析

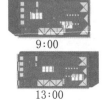
9:00

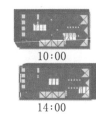
10:00

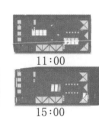
11:00

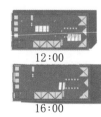
12:00

13:00 14:00 15:00 16:00

方案设计 | Conceptual Design

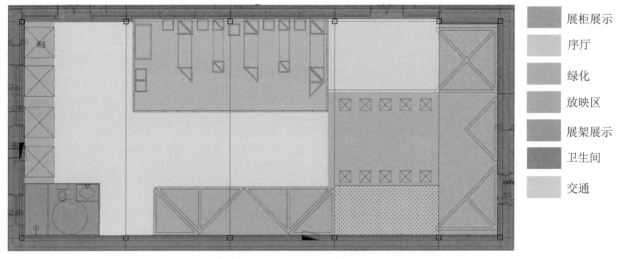

展柜展示
序厅
绿化
放映区
展架展示
卫生间
交通

功能分区图

装配式设计

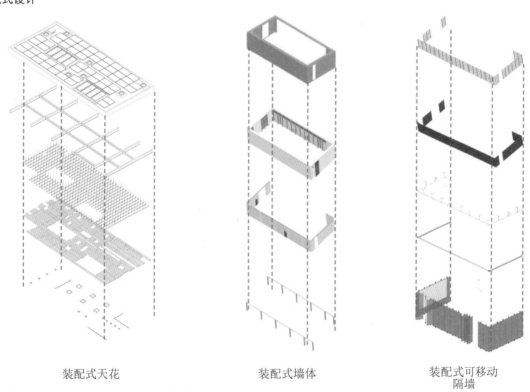

装配式天花   装配式墙体   装配式可移动隔墙

077

材料选取

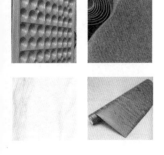

空间划分

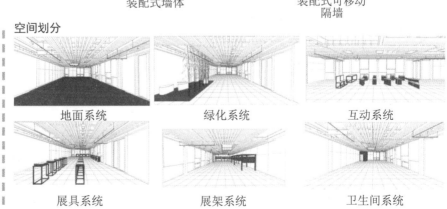

地面系统   绿化系统   互动系统

展具系统   展架系统   卫生间系统

北京建筑大学
Beijing University of Civil Engineering and Architecture

内蒙古阿尔山市明水河镇西口村幸福苑民宿改造
The Renovation of Happiness Garden Homestay in Xikou Village, Mingshui River Town, Alshan City, Inner Mongoliae

「室内设计 6+」2019（第七届）联合毕业设计
"Interior Design 6+"2019(Seventh Year) Joint Graduation Project Event

078

小组成员　马宇萌　乌云塔拉　魏懋榕

改造基本状况 ┃ Basic Overview of Renovation

"阿尔山房"位于阿尔山景区南出口，紧邻内蒙古阿尔山市明水河镇西口村村部楼，属于生态保护区，采用绿色建造技术和装配化施工，与自然和谐共处。

原建筑园区占地 9000 ㎡，每套 7.2×5.1=36.72 ㎡，室内层高 3.0m；双拼 4 栋，3 拼 8 栋，共 16 栋 48 间，其中 4 拼 4 栋，可住 96 ～ 144 人；现将三种房型改造多种功能和样式的民宿，可接待旅友或家庭住宿休闲使用。

从理论上对绿色营造与室内装配式设计进行概念性介绍，初步了解绿色营造——建筑室内装配式设计的理论与实践；了解绿色营造标准、科技、材料、设备等的技术。可以从概念上建立根植于绿色营造的具体生产目标，装配式设计建造为中心的室内设计，需要在实践中进行发展，在当代城乡建设、既有建筑改造更新、室内外装修施工中的应用和发展前景广阔。

由于民宿所在地区，也就是阿尔山市明水河镇西口村的地理位置较为偏僻，在民宿改造的设计与施工中必然要应对其所产生的自然与人文问题。

（1）该地区夏热冬冷，窗户普遍采用了单层玻璃。外窗普遍缺乏遮阳设计，窗户的热工性能差，窗户隔音效果差。

（2）我国寒冷地区建筑墙体保温性能不良，造成了严重的能源浪费，并导致环境污染及室内热环境较差。

（3）我国北方外墙外保温构造的饰面层一般为粘贴外墙饰面砖或抹灰后喷涂外墙涂料。建筑不均匀沉降、结构变形；受太阳辐射应力影响；由于雨水冲刷，外抹灰层或黏结层开裂；外保温板材板缝造成裂缝；由材料原因造成的饰面层开裂，导致在寒冷地区和严寒地区，外饰面开裂带有一定的普遍性。

（4）由于地形偏僻，当地人选择建筑材料更注重可得性和价格廉价性，并不大注重使用的长久度，因而建筑外形受损很严重。

（5）由于我国农村分布范围太广，广大农村居民对农村装配式建筑这一新型建造方式可以说是一无所知，在广大农村居民眼里，他们更愿意相信和接受传统建造方式的质量，而不愿意去进行新的尝试。

（6）装配式建筑的产业链条长、产业分支众多，涉及构件部件、机械设备生产等众多其他产业，而目前我国还是围绕着传统的建筑施工去培养相关专业人才，这无法满足装配式建筑发展对人才的需求，所以专业人才特别稀缺。

（7）装配式建筑的预制构件都是在工厂直接生产，然后运输到现场直接进行安装，这就对道路和运输设备有很高的要求。

（8）阿尔山地区每年霜期长达七个月，在此期间如何有效处理民宿与当地居民的关系成为重中之重。

（9）阿尔山地区归为内蒙古自治区，西与蒙古国接壤，地域文化融合性不足。

项目功能空间的改造方式 | Renovation Ways of Project Functional Space

居住空间中，火炕是北方传统农村住宅中最主要的特色采暖方式，但目前的农村住宅的采暖系统正充分利用当下的先进技术不断进行改进，并越来越朝着低成本、高效率、节能、卫生、美观发展。厨房是农村住宅中的重要组成部分，承担着农民生产、生活及储藏等其他附属功能，是农民生产与生活的联系点。随着农村经济的发展和农民们生活方式的转变，北方地区农村的炊事方式趋于多样化，厨房的形式也随之有所转变。

在农村地区，住宅卫生间的设计一直以来处于尴尬的境地，该地区大部分农村居民仍然使用自己搭建的简易旱厕。然而，近年来开发的沼气技术为解决农村住宅的卫生间入户问题带来了新的希望，农村住宅卫生间入户是农村社会及经济发展的必然趋势，但沼气使用率依旧很低。因此，农村住宅卫生间模块体系的设计应以发展的眼光和现代化的设计方法进行标准化的考量，以符合农村装配式住宅的建造要求。

公共空间的展览空间、休闲空间、娱乐空间、办公空间都可以充分利用装配式展柜与储藏柜，沿墙陈设装配置物柜有效节省空间。提供活动家具与装饰物，模块体系设计多方面、全方位地考虑科技因素对住宅的影响，从而设计出灵活多变、适应性强的模块体系。

装配式建筑在该地区还处于起步建设，标准和技术基础薄弱，市场前景不明朗，培育装配式建筑市场困难。由于地区经济水平、绿色建筑发展基础条件的差异，各方主体的积极性难以有效调动，老百姓认知度低，市场化推动绿色建筑发展的良性机制没有建立，绿色建筑发展仍不平衡。

建筑设计方案初期 ┃ Early Stage Of Architectural Design Scheme

魏懋榕 二拼民宿改造

在阿尔山西口村的民宿设计项目调研中，我们深刻地感受到创意和生活应该是融为一体相辅相成的。整个调研过程中，我们学到了很多。我们体验了当地的食物，欣赏了美丽的风景。那一瞬间我们有更大的欲望把这个项目做好做成，让其他同胞甚至海外朋友也能领会到这样的好风光。这也是四年来为数不多的以小组为单位进行课题的项目。首先，我们有了更多的沟通，努力让对方明白自己的所思所想，使这次实地调研变得更有意义；其次，文献的查询，以及老师还有领导的讲解也给我们带来了更多更好的体验，为后期的改造设计指明了方向，奠定了基础。

居住空间改造空间 ┃ Living Space Renovation Space

设计理念

居住空间是为了满足游客的起居和娱乐结束后的正常生活。我们设计了二拼的标准间和大床房，可以满足不同游客的客房需求。而在室内中，我们主要用木色材质和模数化家具完成此次设计，所有的地面、墙面和天花都是由不同的模块搭建而成。在中期我们已经解决了关于水电走向的技术问题。后期在此基础之上深化并实施。

公共空间改造 ┃ Renovation of Public Space

公共空间

公共空间是为了满足非物质文化遗产的创作与展示的要求，并能让人舒适清新地融入活动之中。

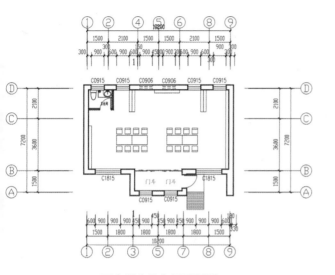

两户型公共空间平面图

设计理念

在这次课题中，传播非物质文化遗产并让游客融入北方真正原始生活变成了最重要的中心。而在这次设计中我们将二拼很小的空间完全打通，一览无余的开阔让狭窄的空间也放大了起来。

在空间中放入了两组桌椅，可以供游客和当地居民更好地交流和教学。在所有的墙面上都设计了展柜更好地展示了当地的刺绣作品，而桌面的空间也足够人们自己动手进行尝试制作。

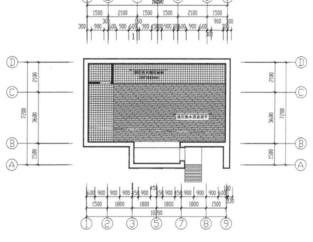

两户型公共空间地铺图

装配化想法

我们参观了许多做绿色装配化的企业，而在聆听了这么多专家和老师的讲解后，我们深刻理解了装配化的便利性和节能性。

对此我们进行了反思，将很多之前的设想完全推翻，并重新着手设计。所有的展柜和家具都采用了300mm×300mm或是它们的倍数化的设备。

除此之外，在选用灯光和桌椅上，我们首先排除异形设备所导致的装配困难，更是去了各种场所真正进入环境深入了解了什么才是真正的装配式。

设计定位 | Design Orientation

　　民宿缘起人们对于品质化、体验化旅游方式的转变。民宿，承载着客户与众不同的期待。也许没有五星酒店的奢华、也许没有多样精彩的娱乐活动。但作为更注重感受化、体验化与价值化的"个性住宿选择"，给住户创造远离城市喧嚣，放松平和、有温度的住宿感受，才是民宿的本质。如今工厂化生产、装配化施工已逐步兴起，建筑装饰已开始从传统操作方法向部品部件生产工厂化、施工现场装配化的方向发展。

　　本次课题将民宿与装配化结合，减少了施工时间与操作难度。

初步设计平面图

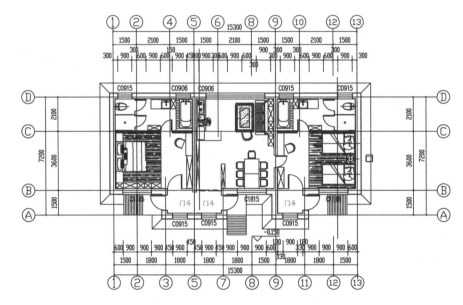

三拼居住空间平面初步设计

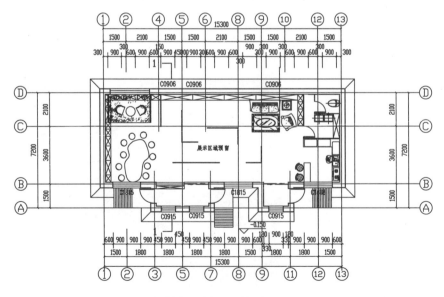

三拼公共空间平面初步设计

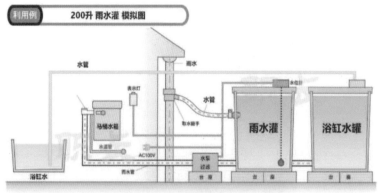

雨水和浴缸水回收再利用系统

　　建筑装饰工厂化生产，就是将建筑装饰的内容模块化、部品化，在工厂里通过规范的工序流程，积极应用自动化、电子化、信息化等先进技术，采用新材料、新工艺、新设备进行工业化生产，并通过运输、现场安装，使得建筑物实现其使用功能，提高建筑物的附加价值，最终满足消费者需求。初步方案将室内进行模数化设计，室内家具全部采用装配式设计，以达到装配化的效果和效率。

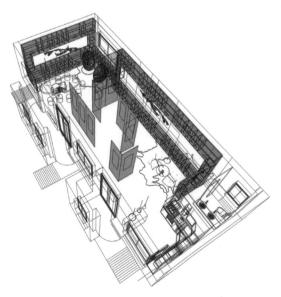

中期检查

乌云塔拉

四拼民宿改造

建筑形制改造 ┃ Architectural form and Structure

　　根据不同的外围护结构采用不同材质与构造，所应用的材质是近几年提倡的环保材质，所应用的构造是近年新兴的内部预制保温构造。屋顶的几何形状具有张力，圆形作为景观中十分有效的焦点性元素，也是作为蒙元文化的一种彰显，如果想让这栋建筑给人以敬仰和探索的感觉，就得保证它本身的就雄浑与神秘。

预制合成树脂瓦
+
倒置式保温屋面

纱网
+
塑钢窗框中空玻璃

清水混凝土砖
+
夹芯外保温复合外墙

石板砖块

藤木外墙框定视野

公共空间改造 ┃ Renovation of Public Space

　　公共空间要义：作为日常活动中心，互补消极空间的流动性，在气候无常，不可出户的时间点发挥核心作用。
　　功能分区要点：品茶区（蒙古砖茶、奶茶）、制作区（手工蘑菇酱、书签等）、展示区（当地木画、编织品等）、休息区（交流休闲）、工作区（劳逸结合）、娱乐区（儿童娱乐与成人娱乐分化）。

　　制作区用以文化传承体验奶茶工艺或者手工艺品。主要沿用绿色营造延伸含义——仿生，装配仿真门窗，营造静谧的氛围，符合慢的主题。

　　娱乐区的桌游设施采用装配式可拆卸移动结构，尽量去处理好旺季与体验者的互动性，以及淡季与当地村民的关系。东面墙体作为进入时的主要装饰面，运用纹饰浮雕。

084

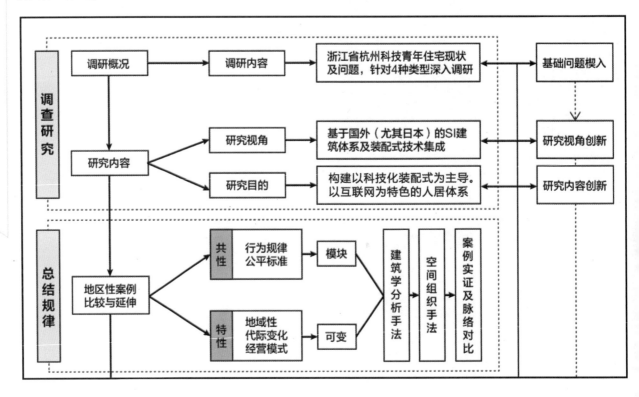

居住空间改造 ｜ Renovation of Living Space

住宿空间：家庭套间与朋友标间，以满足人们对舒适与睡眠的渴求。

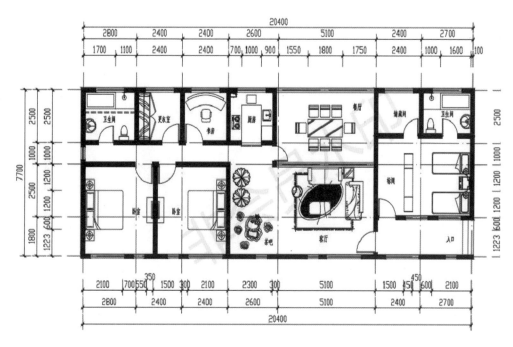

与穿行的消极空间互补的住宿空间需要满足驻留感——休憩要舒适。在当地绿色天然木的清新中，躺在装配的蒙古包状帷帐里，望着俄式彩绘窗帘随风摇曳，在文化遗产的馈赠里尽情享受，疲累感在睡眠中得到缓解。

早上醒来如厕或者洗浴过后，看着装配大镜面装饰隔断的绿色山水画，引导人们走向外面欣赏阿尔山如诗如画的大好山水。

「室内设计 6+」2019（第七届）联合毕业设计
"Interior Design 6+"2019(Seventh Year) Joint Graduation Project Event

中国国际进口博览会展示设计 1
Exhibition Design of China International Import Expo 1

南京艺术学院
Nanjing University of the Arts

小组成员
陈 涛
叶子萱
李艺蓓
李 赞

企业调研 ┃ Corporate Research

科技创造美好生活
——"Leaps By Bayer"

创新是企业的核心精神。

拜耳的愿景是通过创新解决方案，在健康与营养领域，为客户和社会创造附加值。拜耳的业务活动基于强大的研发能力和全球开放式创新网络。

拜耳主要产品

设计理念的形成 ┃ Formation of Design Concept

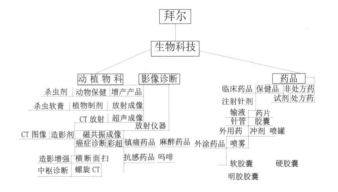

观众分析 ┃ Audience Analysis

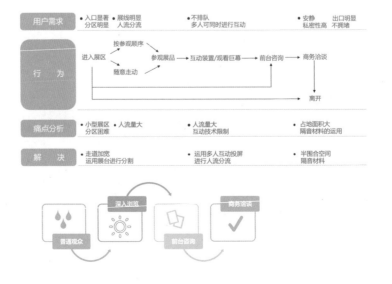

装配化应用 ┃ Assembly Application

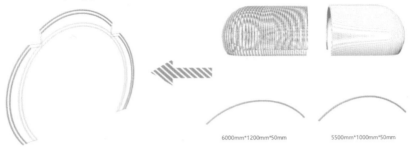

6000mm*1200mm*50mm   5500mm*1000mm*50mm

企业调研 ┃ Corporate Research

构建万物互联的智能世界
宇宙遨游 万物生长

　　华为的追求是实现客户的梦想。
　　华为将聚焦客户关注的挑战和压力，提供有竞争力的通信解决方案和服务，持续为客户创造最大价值。
　　为了成为世界一流的设备供应商，华为不断加强科研。在独立自主的基础上，开放合作地发展领先的核心技术体系，用卓越的产品自立于世界通信列强之林。

营造独特性格

建立可识别性

提高可达性+可视性

价值倍增布局

绿色装配化搭建

设计理念的形成 ┃ Formation of Design Concept

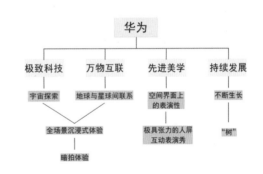

流线分析 ┃ Streamline Analysis

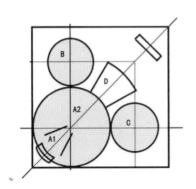

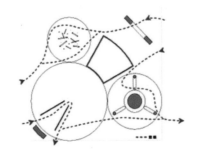

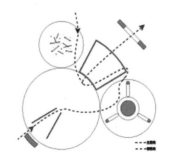

南京艺术学院
Nanjing University of the Arts

中国国际进口博览会展示设计2
Exhibition Design of China International Import Expo 2

「室内设计 6+」2019（第七届）联合毕业设计
"Interior Design 6+"2019(Seventh Year) Joint Graduation Project Event

国家调研 ┃ State Survecy

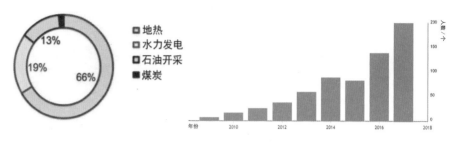

- ☐ 地热
- ☐ 水力发电
- ☐ 石油开采
- ■ 煤炭

13%
19%
66%

冰岛能源使用占比

冰岛近年来旅游人次变化

　　冰岛作为世界上首个零碳排放的数据中心，率先迈向零碳的国家，在当今资源紧张的局势下，无疑是值得参考学习的。

　　冰岛独特的自然风光每年都吸引了大量的游客。旅游业已成为冰岛的支柱产业之一，其产业效益在 2010—2018 年中迅速增长。冰岛正在逐年加大和中国的贸易往来。

　　中国政府正大力寻求推广清洁能源使用并降低污染，进博会将会是冰岛向中国大力推出清洁能源的技术的平台，冰岛馆也将再生能源技术作为进博会国家馆的主要展示方向。

首届进博会中国馆："共羽华平"
设计理念：斗拱、飞檐等传统建筑元素给人以开放、大气之感
1500 ㎡，8 个展馆，其中 3 个为港澳台
传递开放、融通、共享理念，蕴含构建人类命运共同体的美好期许，
也昭示着中国开放的大门不会关闭，只会越开越大。

中国馆概念演变 ┃ Evolution of the Concept of the Chinese Pavilion

$Z = \cos(x^2 + y^2) \cdot e^{-x^2 - y^2}$

小组成员
王紫荆
朱一丰
王紫薇
高榕泽

人流量及流线分析 | Analysis of Human Flow and Streamline

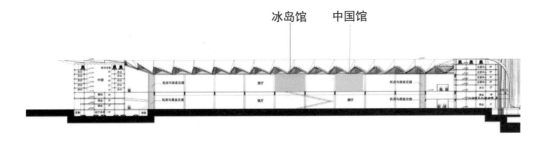

冰岛馆　中国馆

单日国家馆参观人数人数变化趋势

10 万

5 万

8时　10时　12时　14时　16时　18时

普通观众　| 5～15min |　参观 15～30 个

贵宾领导　| 15～20min |　参观 5～12 个

中国馆　| 197～3545 人次

主宾国　| 34～775 人次

其他国家　| 16～387 人次

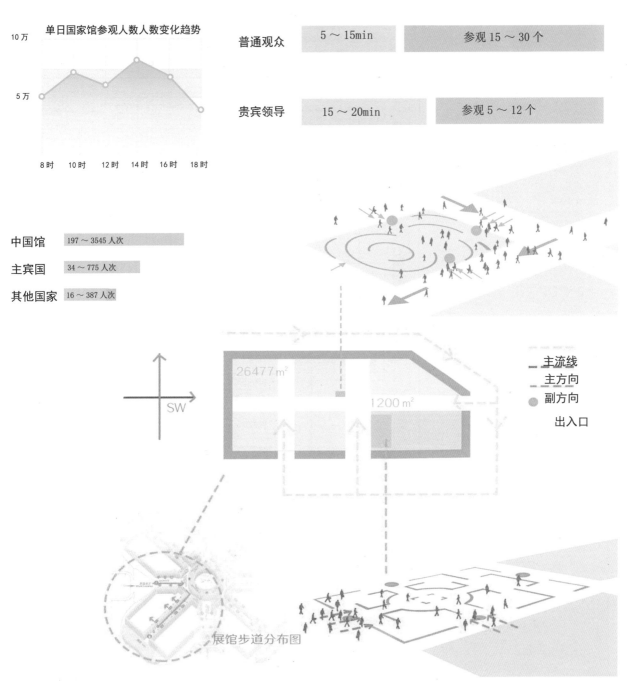

26477 m²

1200 m²

SW

—— 主流线
—— 主方向
● 副方向
出入口

展馆步道分布图

浙江工业大学
Zhejiang University of Technology

基于一体化设计与装配化建造的阿里巴巴青年公寓室内设计研究1

Interior Design Research of Alibaba Youth Apartment Based on Integrated Design and Assembly Construction 1

『室内设计 6+』2019（第七届）联合毕业设计

"Interior Design 6+"2019(Seventh Year) Joint Graduation Project Event

090

小组成员
沈令逸
张毅津

概念提出 ┃ Proposed Concept

目前的居住形式多为
分室合租

您目前的居住形式？

A 分室合租 ████████████████ 84.21%
B 同室合租 ██ 5.26%
C 独居整套房屋 ████ 10.53%

➡ 将个人生活空间独立
可共享功能外置

大部分人会选择
21 ～ 60 ㎡的生活空间

在您目前的经济条件下，您会租用
多大面积的房子？

A 20㎡ 以下 ████ 21.05%
B 21 ～ 60㎡ ████████ 50.74%
C 60 ～ 90m² █ 5.26%
D 90m² 以上 ████ 22.95%

➡ 提高空间利用度，将
每个空间面积压缩

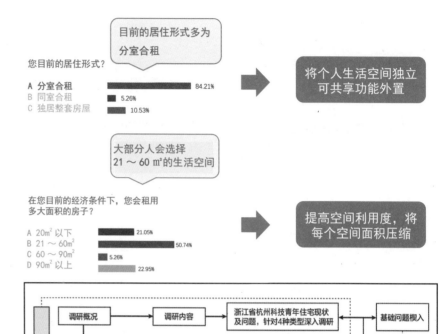

生活方式转变 ┃ Lifestyle Change

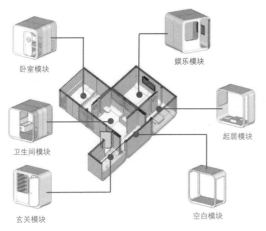

卧室模块
娱乐模块
起居模块
卫生间模块
玄关模块
空白模块

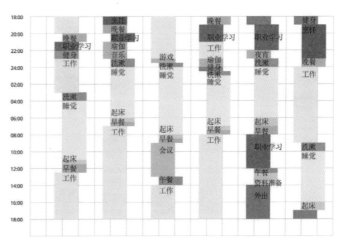

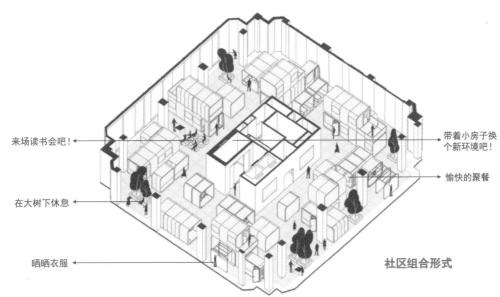

来场读书会吧！

带着小房子换个新环境吧！

愉快的聚餐

在大树下休息

晒晒衣服

社区组合形式

对共享住宅的看法

您认为社区中可以增加的公共功能有（多选）？

A 学习交流空间
B 休闲娱乐空间
C 健身空间
D 商业空间
E 餐饮空间
F 其他

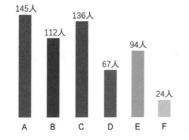

145人 A
112人 B
136人 C
67人 D
94人 E
24人 F

您对青年公寓、青年共享社区这类青年人聚居的社区是否有居住意愿？

A 有
B 无

愿意入住青年共享社区的人数较多

26.32%
73.68%

■ A
■ B

科技化住宅的吸引力较强

未来什么样的住宅会更加吸引您？

A 可与邻居共享空间的住宅
B 科技化住宅
C 就现在这样的
D 其他

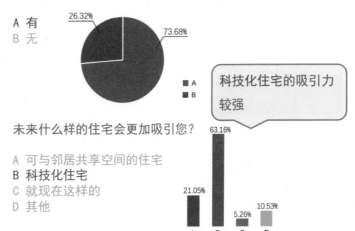

21.05% A
63.16% B
5.26% C
10.53% D

091

浙江工业大学
Zhejiang University of Technology

基于一体设计与装配化建造的阿里巴巴青年公寓室内设计研究 2
Interior Design Research of Alibaba Youth Apartment Based on Integrated Design and Assembly Construction 2

「室内设计 6+」2019（第七届）联合毕业设计
"Interior Design 6+"2019(Seventh Year) Joint Graduation Project Event

小组成员
范诗意
陈 格

设计框架提出 IDesign framework

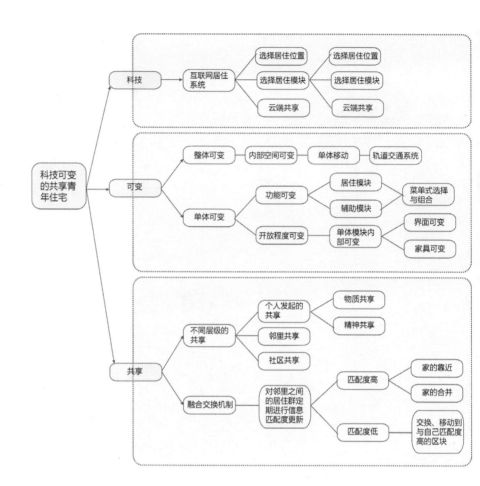

平面网格系统 ┃ Plane Grid System

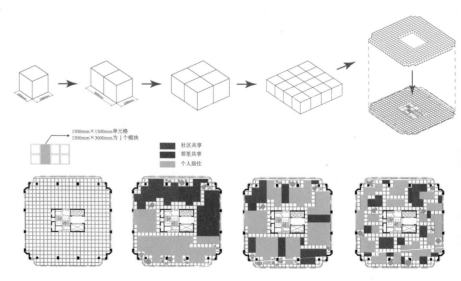

在每层的平面布局上，我们将原本相对独立的个人居住、邻里共享以及社区共享进行打散再打散，把邻里共享和社区共享穿插在个人居住环境中，使居住者们能主动地去参与共享的部分。

功能模块划分 ┃ Functional Module Division

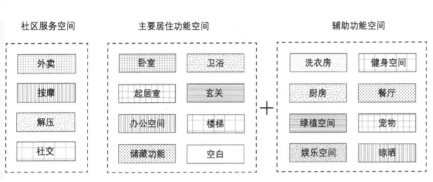

社区服务空间

| 外卖 |
| 按摩 |
| 解压 |
| 社交 |

主要居住功能空间

卧室	卫浴
起居室	玄关
办公空间	楼梯
储藏功能	空白

辅助功能空间

洗衣房	健身空间
厨房	餐厅
绿植空间	宠物
娱乐空间	晾晒

根据前期两次调研的结果，我们首先将模块按照不同的功能进行了划分，主要分为满足目标人群基本生活的居住功能模块、满足目标人群个人特殊需求的辅助公共功能模块以及满足目标人群这类人所特有的社区服务空间。然后在三种大类下对模块进行细分。每个人的空间由主要的居住功能模块和辅助公共模块组成。

模块界面选择 ┃ Module Interface Selection

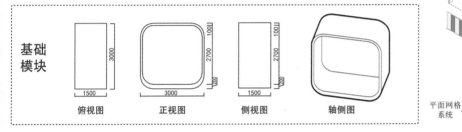

基础
模块

| 俯视图 | 正视图 | 侧视图 | 轴侧图 |

用户可以根据自己对居住环境的需求进行菜单式的选择，最终组合成属于自己的家。同时，考虑到模块与模块之间的连接，我们对每个模块的六个界面也进行了考虑。在用户选择完他所有的模块后，系统将为其合理地进行模块间的组织，提供给用户舒适的居住环境及交通流线。

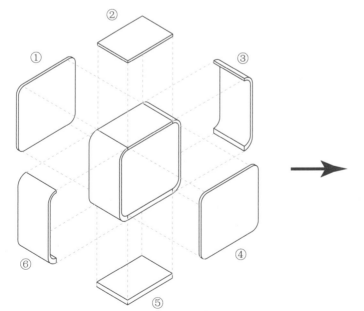

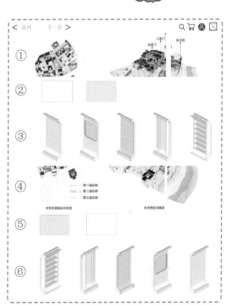

模块组合

建筑内部
升降机

平面网格
系统

联合指导　服务需求

热点命题　纷显特色

答辩评审

『室内设计6+』2019（第七届）联合毕业设计

"Interior Design 6+"2019(Seventh Year) Joint
Graduation Project Event

春播夏收

落实课题目标 调动专业手段 综合解决设计问题

讲述图示艺术语言 展现团队风采

高　　校：同济大学

College: Tongji University

学　　生：黄曼姝、詹强、王萌、周宽

Students: Huang Manshu, Zhan Qiang, Wang Meng, Zhou Kuan

指导教师：左琰、林怡

Instructors: Zuo Yan, Lin Yi

课题分数：85

Subject Scores: 85

小户型租赁住宅装配化设计——深圳『城中村』住宅改造

Assembly Design of The Small Rental Housing—Housing Renovation of "Urban Villages" in Shenzhen

黄曼姝

詹强

王萌

周宽

学生感悟

Student's Thought

　　通过这一阶段的努力，此次毕业设计终于告一段落，这意味着大学生活即将结束。在大学阶段，我在学习上和思想上都受益匪浅，这除了自身的努力外，与各位老师、各位同学的关心、支持和鼓励是分不开的。

　　在这次小户型租赁住宅的设计研究过程中，我受益匪浅。感谢老师们的认真指导，自己接触室内设计方面内容时间不长，经验不足，老师们耐心的指导让我对室内设计的流程有了比较完整的认识。在设计中期，我们小组在设计概念选取方面遇到了瓶颈，感谢老师们细心的梳理，使我们明确了设计方向。

　　本次小户型租赁住宅课题设计作为本科最后一次课程设计，不仅是对之前所学知识的检验、梳理与补充，更使我从中学习到很多。从城市到建筑、再到室内，每一个尺度都需要方方面面的考量，关注到租户、运营商、社区居民等多种人群的需求。最后诚挚感谢老师们的耐心指导以及小伙伴们的合作互助！

　　作为本科阶段的最后一次设计，本次课题涉及范围广、政策联系强，具有一定的挑战性。通过对本科以来所学的内容进行"装配"，我们对于这次的装配化改造设计有了更深的理解。对市场、政策和建造本身，也有了更深刻的认识。最后感谢老师的倾情指导和同学的热心帮助！

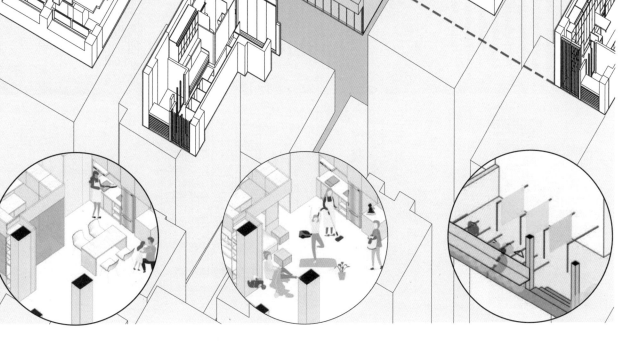

因地制宜的遮阳策略 ︳ Shading Strategies Tailored to Local Conditionssummary

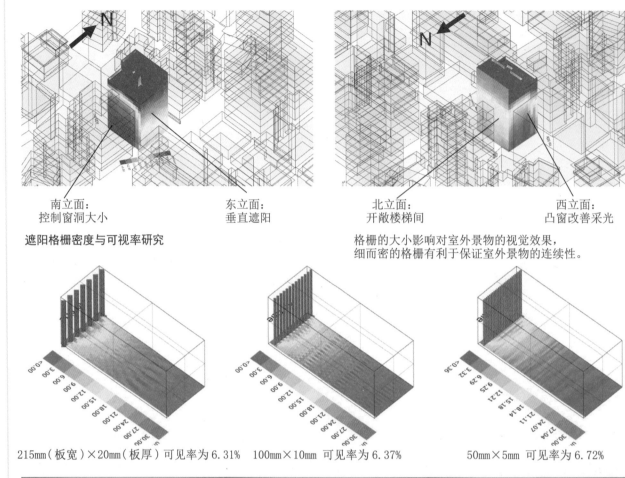

南立面：
控制窗洞大小

东立面：
垂直遮阳

北立面：
开敞楼梯间

西立面：
凸窗改善采光

遮阳格栅密度与可视率研究

格栅的大小影响对室外景物的视觉效果，
细而密的格栅有利于保证室外景物的连续性。

215mm（板宽）×20mm（板厚）可见率为6.31%　　100mm×10mm 可见率为6.37%　　　50mm×5mm 可见率为6.72%

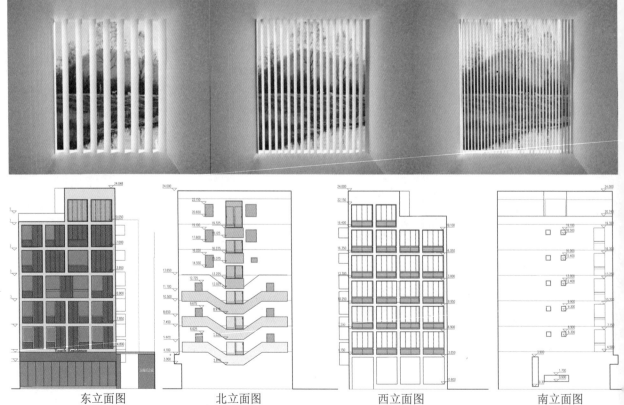

东立面图　　　　　北立面图　　　　　西立面图　　　　　南立面图

遮阳格栅与落地窗结合 | Combination Between Shading Grille and Floor-to-Ceiling Windows

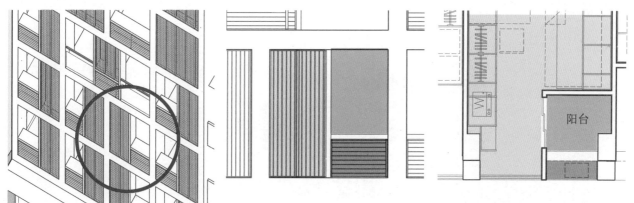

东立面

落地窗　空调机位

落地窗　空调机位

西立面

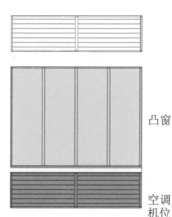

凸窗

空调机位

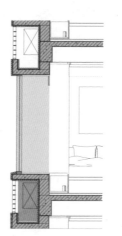

外立面与落地窗结合，空调机位隐藏在阳台窗台下，不影响室内使用空间。

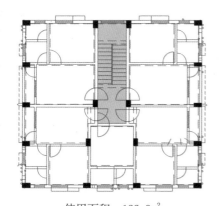

使用面积：122.9m²
公共面积：18.9m²

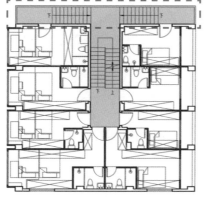

使用面积：127.1m²
公共面积：30.1m²

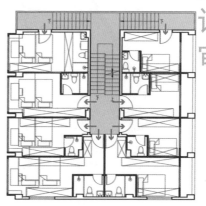

改善入户空间拥挤状态

项目	数量	单价	总价/元
钢结构	300kg	8元/kg	2400
扶手栏杆	14 m²	150元/m²	2100
踏步及平台	12 m²	150元/m²	1800
人工	12 m²	200元/m²	2400
合计			8700

使用面积增加了 4.2m²（3.4%），公共空间增加了 11.2m²（59.1%）。

三个楼梯间的造价约为 26100 元，按照 200 元 /（m²·月），约 10 个月回收成本。

门厅效果图

利用阶梯的平台和模数化的凳子围合、塑造适应不同功能的空间形式。

舞台讲堂　　　阅读休憩　　　聚会交流

根据90后群体个性张扬、追求新鲜感的特点，掌握青年群体对各种活动的喜爱程度作为功能参考。同时考虑到建筑本身特点，补充了快递、外卖的寄存和洗衣烘干等功能，强化在小面积下的实用性特点。

在整体风格上，采取大面积灰黄的主色调，以营造空间温馨、舒适的整体氛围。材料上以木制、树脂为主，配合自流平地面呈现出亲和而朴实的独特效果。

底层公区通过利用原本自身的层高优势拓展了夹层空间，使得总面积连带夹层达到了约110m²，均摊到每户足有3m²，与目前市场上的长租公寓公区面积情况相比也属前列。

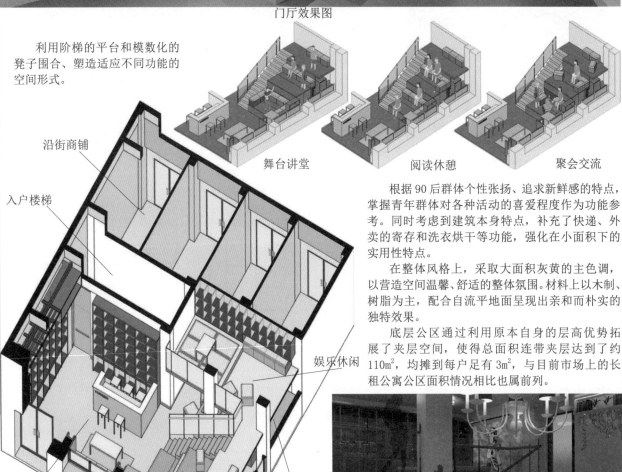

沿街商铺

入户楼梯

娱乐休闲

餐厨水吧

门厅接待

阶梯空间　　　洗衣烘干

阶梯效果图

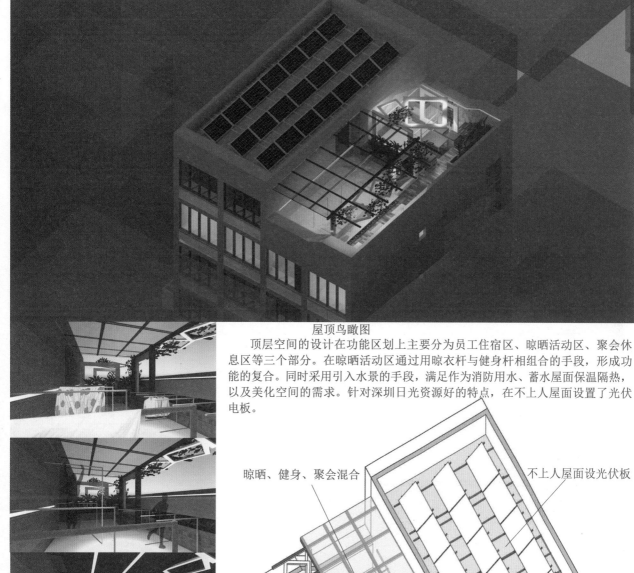

屋顶鸟瞰图

顶层空间的设计在功能区划上主要分为员工住宿区、晾晒活动区、聚会休息区等三个部分。在晾晒活动区通过用晾衣杆与健身杆相组合的手段，形成功能的复合。同时采用引入水景的手段，满足作为消防用水、蓄水屋面保温隔热，以及美化空间的需求。针对深圳日光资源好的特点，在不上人屋面设置了光伏电板。

晾晒、健身、聚会混合

不上人屋面设光伏板

休息聚会

员工宿舍

景观水池间蓄水

A1：单人独居户型

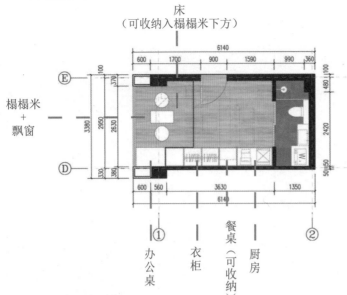

床
（可收纳入榻榻米下方）

榻榻米
+
飘窗

办公桌　衣柜　餐桌（可收纳）　厨房

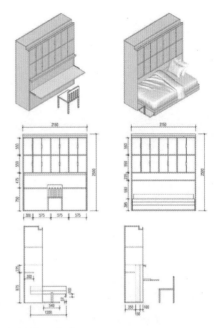

床（带桌）

B2：情侣／夫妻同居户型

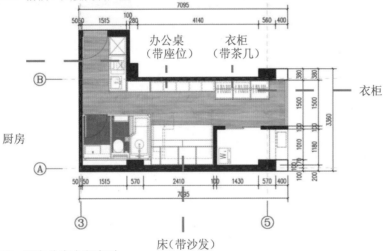

办公桌
（带座位）　衣柜（带茶几）

厨房

衣柜

床（带沙发）

C2：朋友分室合租户型

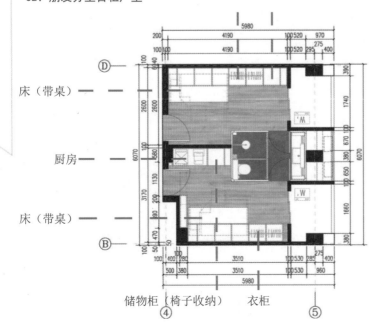

床（带桌）

厨房

床（带桌）

储物柜（椅子收纳）　衣柜

部件名称	长/mm	宽/mm	厚/mm	数量/个
侧板	2475	600	25	2
顶板	2150	455	25	1
背板	2500	2150	25	1
底板	2100	600	25	1
顶部灯槽背板	2150	100	25	1
顶部灯槽底板	2150	145	25	1
顶部灯槽前板	2150	50	10	1
柜门	575	550	25	6
柜门-2	575	300	25	2
内底板	600	575	25	3
内底板-2	600	300	25	1
内底板-3	2100	600	25	1
内侧板	1225	600	25	3
置物板	2100	300	25	1
床板	2080	1200	100	1
桌板	2080	540	30	1

柜顶灯槽

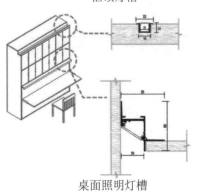

桌面照明灯槽

「室内设计 6+」2019（第七届）联合毕业设计
"Interior Design 6+"2019(Seventh Year) Joint Graduation Project Event

照明设计——全开模式

工作重点照明模式

睡前阅读模式

睡眠模式

家庭户型设计 ┃ House Type Design of Family

5层
使用面积：
31.2 ㎡

家庭成员：夫妻 + 婴幼儿
户型类型：一室户
户型特点：开放式

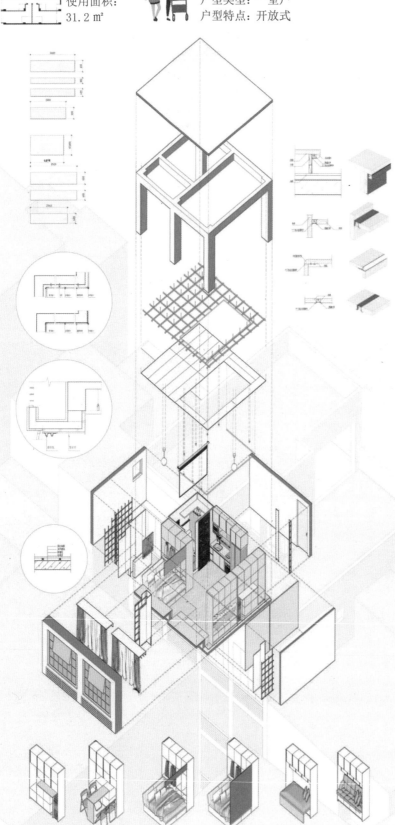

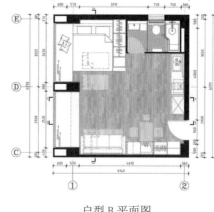

户型 B 平面图

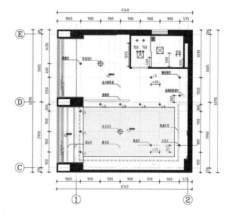

户型 B 平顶图

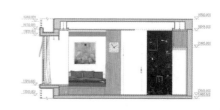

户型 B 剖面图 1

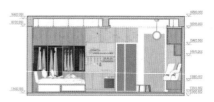

户型 B 剖面图 2

户型 B 剖面图 3

『室内设计 6+』2019（第七届）联合毕业设计
"Interior Design 6+"2019(Seventh Year) Joint Graduation Project Event

户型 B 效果图

晚餐　　　　观影

结构爆炸分析

1. 实木平开门　　4.SMC 石纹壁板　7. 塑钢框架
2.SMC 防水底盘　5. 吊顶板　　　　8. 卫浴器具
3. 玻璃移门　　　6. 顶面板

透明玻璃　黑色塑钢　SMC板材　陶瓷塑脂　复合木板

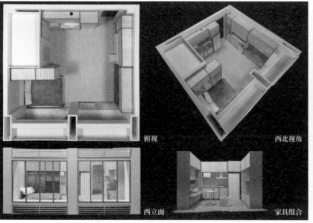

俯视　　　　西北视角

西立面　　　家具组合

105

俯视

东立面　　　立面

陈卫新

该方案整体上很接近实际设计的项目，各方面考虑得非常周详。37栋的该处房子客观来看是城中村的老房子，对此提出几点意见：

（1）方案中首先应当考虑其安全性、功能性和需求性。其次方案中关于建筑外立面和室内两者之间的协同性尚存在一些问题。例如南立面、西立面等虽然有不同的处理方法，且单个看每个立面都是不错的。但各个立面之间的关系不是割裂的，可以再深入些处理。

（2）在户外入口处要考虑用户的体验感与方案实现的可能性。从目前户外的入口情况来看，未来加建楼梯的可实现度要慎重考虑以什么方式呈现出来，以及可能随之产生的噪声等问题如何解决。在室内部分，由于改造后上面有一个很大的消防水箱，屋顶加建和荷载的问题也需要注意。

（3）另外，体育和休闲互动可变性的目的从目前来看很难实现。主要原因是体育健身设施本身存在一个安全性的规范要求。除此之外，正常的安全安装固定都要考虑。

（4）家具部分，细节做得很细致。从建筑室内装配化的角度来看，该设计方案中的家具部分包括厨房和卫浴环节都考虑得很周全，尤其是考虑到了下水管道的问题。这一点是极其重要的，因为每个楼层的套型不一样，所以在设备设施方面也可能面临不同的问题。

（5）隔墙之间和公共区噪声等部分还需要进一步的深化设计。

徐敏

该方案从完整的调研到分析，再到设计环节的整体过程汇报比较完整。其设计过程包含从室外—户型—室内，再到排水管道。在方案阐述中，逻辑性也非常强。其中在户型分析和室内装配化主题上，十分契合这次毕业设计的主题。但还是存在以下几点问题：

（1）虽然汇报的分析点有很多，但是并没有就某个点非常深入细化的。比如说在装配化和调研方面。

（2）关于整个户型的分析，涉及从十几平方米到三十几平方米左右的户型之间的关系，那么户型与90后年龄段的适应性是怎样的？是接近30岁的人还是20岁的人？需要考虑这个空间对他们生活的舒适度是否足够。该组的同学在室内功能方面，从安全性、功能性和舒适度上做了详细分析，但是这些分析只是考虑了感官部分舒适性，还需要考虑在精神方面、职业方面的舒适性。

（3）由于室内空间非常狭小，需要在公共空间增加大量辅助空间，提高室内空间窄小的公共互相交互的使用率，也就是要发展共享空间。

（4）另外，厨房的设计没有封闭。按照国家规定，厨房是必须封闭的。

（5）厨房很小，但是卫生间却很大。因此对于卫生间舒适度的要求高这一认识到底是不是对90后理想价值的全面认识呢？这个问题有待探讨。

刘杰

通过PPT呈现出来的设计的逻辑性工作方法值得学习。设计方案中解决的一个最大的问题就是如何能够最大化利用空间。尤其是方案中设计者把每一层平面都做得非常详细，同时家具配置和装配化实施的空间（例如厨房和洗手间）也都进行了细化，使3～4层装配化率达到了88%。设计除了理念的落地实施以外，它还达到了相当的细致设计程度。由于考虑到越高楼层的光照条件越好，方案中将阳光分配给家庭房是个合理的选择。另外小户型设计中过多注重安全性，就会少了舒适性，但该方案通过装配式的形式很好地解决了舒适性的问题。方案根据居住者不同的活动需要，赋予空间更加灵活性。同时方案也具有社会层面的意义，例如设计师给予城市中最辛苦的人群以最大的关怀，令其生存方式和生活态度更加有尊严感。

The scheme is generally close to a real design project with very meticulous considerations in various aspects. The house of building No.37 shall be an old building in urban village objectively speaking, and the following opinions are given:

(1) The safety, function and demand shall be considered in the scheme. Secondly, the scheme contains some coordination problems of building external façade and interior. For example, the southern façade and western façade are handled in different ways, and seem good separately. But, the relationship between the facades shall not be broken, and can be more thoroughly handled.

(2) The user's experience and scheme realization probability shall be considered for the outdoor entrance. It can be seen from the current outdoor entrance condition that the feasibility of staircase construction in the future, the way to realize, and the way to solve the potential related noise problem shall be meticulously considered. In indoor area, since the large firefighting water tank is set above after renovation, the issue of roof additional construction and load shall be considered.

(3) Moreover, it can be seen that it is difficult to fulfill the purpose of sports and leisure interaction and changeability currently. The main reason is that the sports and fitness facilities inherently bring the safety standard requirements. In addition, the normal safety installation and fixture shall be also considered.

(4) The furniture details are delicately handled. From the perspective of building indoor assembly, the furniture of the design scheme, including kitchen, toilet and bathroom, is comprehensively considered. The drain pipe is especially taken into consideration, which is of great importance. Because the house type in every floor is different, the equipment and facilities may face different problems.

(5) Further in-depth design shall be undertaken for the areas between the partition wall and public area noise, etc.

The scheme of the team, from complete survey, analysis to overall process of design, is relatively comprehensively reported. The design process covers outdoor – house type – indoor and drain pipe. The scheme explanation is highly logical. Especially the house type analysis and indoor assembly perfectly match the theme of the graduation design. However, the problems are as follows:

(1) Though many analyses are given in the report, no point is thoroughly refined, such as assembly and survey.

(2) The analysis of the house type involves the relationship among the house types from more than ten square meters to more than thirty square meters. But what is the applicability of the house types in the post-90s generation? Is it friendlier to those about 30 years old or 20 years old? It shall be considered that whether the space is sufficient to make their life comfortable. The students in this team analyze the indoor functions from the aspects of safety, function and comfortableness. However, the analyses only consider the sensual comfortableness. The spiritual and occupational comfortableness shall also be taken into consideration.

(3) Because the indoor space is very small, large auxiliary space shall be arranged in public space and the use ratio of public interaction in the place with small indoor space shall be increased.

(4) Moreover, the kitchen is not in closed design. The national provisions prescribe that the kitchen must be closed.

(5) The kitchen is very small but the toilet is very big. So is the understanding that the post-90s generation has high requirement for toilet comfortableness a comprehensive idea of the ideal value held by the post-90s generation? This question is to be discussed.

The others can learn from the logical work method of design represented by PPT. The design scheme solves the greatest issue of maximum utilization of space. The scheme designer especially handles every floor plans in details, and refines the furniture configuration and assembly execution space (i.e. kitchen and toilet) to achieve assembly rate of 88% in F3-4. The design not only contains the realization of the idea, but also achieves high design refinement level. With the consideration that the higher floors receive better lighting, the scheme makes the reasonable choice to assign the sunshine to the family unit. In addition, the small house type design pays excessive attention to safety, which thus reduces comfortableness. But the scheme solves the comfortableness issue through assembly form. The scheme develops more flexibility of space according to different activity demands of the inhabitants. Meanwhile, the scheme is also of social significance. For example, the designer gives the greatest care to the most hard-working group in the city to bring more sense of dignity to their way of survival and life attitude.

高　　校：华南理工大学

College: South China University of Technology

学　　生：黄皓庭、典超华

Students: Huang Haoting, Dian Chaohua

指导教师：薛颖、谢冠一

Instructors: Xue Ying, Xie Guanyi

课题分数：85

Subject Scores: 85

华南理工大学 A4 教学楼一层更新改造设计

The First Floor Renovation Design of A4 Teaching Building in South China University of Technology

黄皓庭

典超华

学生感悟
Student's Thought

　　本次毕业设计得以顺利完成，除了感谢谢冠一老师、薛颖老师的鱼渔双授之外，还要感谢我的组员典超华。从毕业设计开题之际，到调研、汇报、方案等阶段，我们有过意见分歧，有过支持陪伴，感谢你在每个挑灯夜战的夜晚与我并肩奋战。

　　我在完成作品的过程中学会更好地倾听队友的想法，把我们的优势结合起来，更好地完成作品，从分析课题，到寻找案例，分析案例，最后完成设计，过程中遇到过一些困难，但是在克服难题的过程中我也受到了很大的启发，很感谢这次的经历，带给我很多不一样的感受。

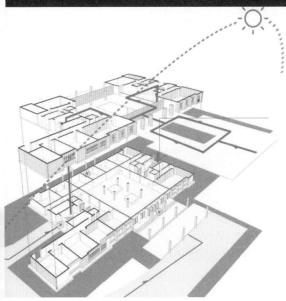

现象分析 | Phenomenon Analysis

在调研中发现，该场地一共有 10 个出入口，但基本保持关闭状态。

若出入口常年关闭，是否意味着该场地的使用率较低？但根据调研得知，该教学楼课室的使用率较高。而大部分时间里，学生和老师却并不能直接从教学楼的首层进入课室，而是要穿越教学楼平台进入二楼，再从二楼走到首层，交通流线极为复杂。

问题分析 | Problem Analysis

造成上述现象的主要原因在于该场地的课室是供全校师生上课的场所，而展厅则是一个专供设计学院使用的场所。由于两种空间类型其归属方的不同，为了方便场所的管理，故展厅在没有展览的时候将首层出入口封闭。

通过调研该展厅我们发现，展厅大部分时间都是呈现闲置的状态，故展厅在大部分时间都是处于封闭状态。

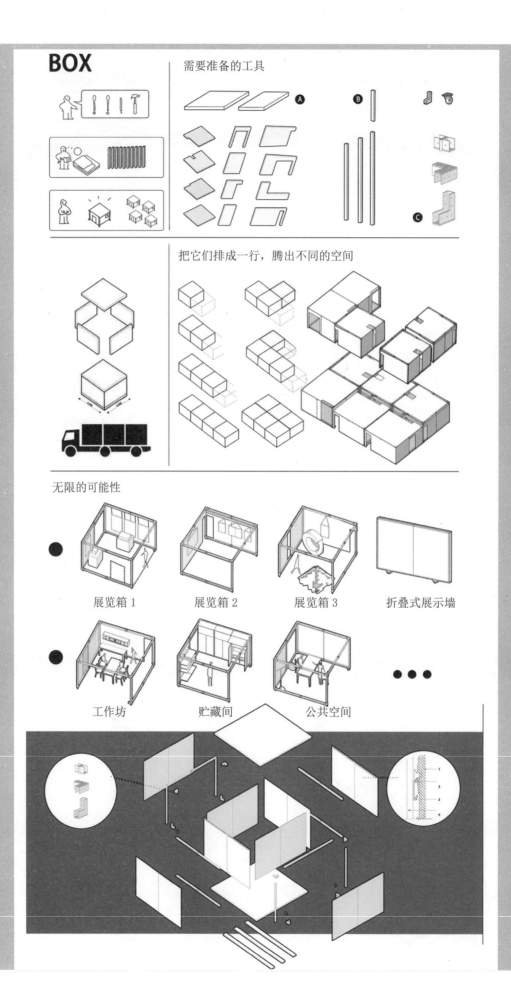

BOX

需要准备的工具

把它们排成一行，腾出不同的空间

无限的可能性

展览箱 1   展览箱 2   展览箱 3   折叠式展示墙

工作坊   贮藏间   公共空间

展厅模式一

宣讲模式

展厅模式二

工作坊模式一
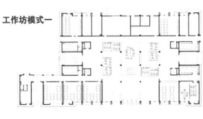

展厅模式三

工作坊模式二
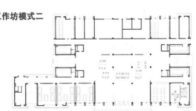

陈卫新

关于绿色营造，装配化的主题性在方案汇报里面没有把它明确提出来，方案后部分环节可以有更深入的亮点讲解，使听者更容易理解。根据汇报的成果来看，通过从盒子的侧面设计，盒子（4m×4m）的尺寸应用，包括盒子边缘的变化过程就像方案之前提及的三种形式：盒子与盒子、盒子与展墙、展墙与展墙。实际都是在盒子当中边缘变化形成的一个成果。在空间利用层面，包括上进性的问题，中间的共享区和工作坊之间有着怎样的关系，还可以详细说明。在营造这几种盒子的组变方式上，需要关联到装配化设计的角度。

林怡

该组设计方案在设计之初，对其校园内现有建筑内部公共空间的使用状态进行了较充分的分析，挖掘出设计对象的关键问题"孤独"和"自闭"。这是个生动的比喻，随之而来的设计切入点也非常好。设计者采用了装配式的盒子结构作为设计手段对空间进行了设计再造，轻质的框架板材结构为空间分隔提供了充分的灵活性与可变性，并创造了相对丰富和又相互独立的不同功能子空间。如果能在具体设计细节和手法上进一步思考和探索，如盒子的材质和色彩是不是可以有更多不同的选择，还有什么样的手段和方法能够使得空间更为开放和自由，促进更多的沟通和交流，相信这能够对原初的问题有更好的回应。

吴晓燕

该组方案设想很好，但是内容感觉不够完整，还可以继续深入地细化。本次活动中的所有设计方案，无论在前期调研，还是功能和流线分析上都做得相当充分。虽然课题都做得很扎实，但最后呈现出来的整体视觉美感不足，在此提一点建议。"室内设计 6+"活动可以借鉴中央美院一年级的毕业设计导师团队的组建方式，每个学生的导师包括本专业的第一导师和其他绘画或者视觉传达等专业的第二导师，通过综合性学科的共同辅导，所呈现的毕业设计作品无论从作品的深度还是美观上，都会体现出高度的专业性和审美性！

In terms of the green construction , the theme of assembly is not explicitly put forward in the scheme report. The latter part of the scheme can be explained in more in-depth highlights, so that the listeners can understand it more easily. According to the reported results, through the side design of the box, the size application of the box 4M×4M, including the change process of the edge of the box, is like the three forms mentioned before : box and box, box and exhibition wall, exhibition wall and exhibition wall. It's all a result of the change in the edges of the box. At the level of space utilization, including aspirational issues, the relationship between the shared area and the workshop in the middle can be explained in detail. In the way of creating these types of boxes, it is necessary to relate to the perspective of assembly design.

At the beginning of the design, the design of the group made a full analysis of the use of the public space inside the existing building on the campus, and unearthed the key issues of the design object" lonely" and" autistic". This is a vivid metaphor, and the design entry point that comes with it is also very good. The designer adopts the prefabricated box structure as the design method to design the space again. The lightweight frame and plate structure provides sufficient flexibility and variability for space separation and creates relatively rich and independent sub-spaces with different functions. If we can further think and explore in the specific design details and techniques, such as whether there can be more different choices for the material and color of the box, and what means and methods can make the space more open and free, and promote more communication and exchange, we believe that this can better respond to the original problem.

Schemes of this group is well conceived, but the content is incomplete and can be further refined. All the design schemes in this activity, whether in the preliminary investigation, or in the function and streamline analysis, have been completed quite adequately. Although the subject is very solid, the overall visual aesthetics presented at the end is not enough. I would like to make some suggestions here. The activity of" interior design 6+" can draw lessons from the formation of the first-grade graduation design instructor team of the Central Academy of Fine Arts. Each student's tutor includes the first tutor of the major and other second tutors of painting or visual communication and so on. Through the joint guidance of comprehensive disciplines, the graduation design works presented will reflect the high degree of professionalism and aesthetics in terms of the depth and beauty of the works!

答辩评审

『室内设计 6+』2019（第七届）联合毕业设计
"Interior Design 6+"2019(Seventh Year) Joint Graduation Project Event

雄安设计中心室内外环境更新设计
Indoor and Outdoor Renewal Design of Xiong'an Design Center

高　　校：哈尔滨工业大学
College: Harbin Institute of Technology
学　　生：林慧颖、李博扬、秦卫杰、王祺叡
Students: Lin Huiying, Li Boyang, Qin Weijie, Wang Qirui
指导教师：刘杰、马辉
Instructors: Liu Jie, Ma Hui
课题分数：85
Subject Scores: 85

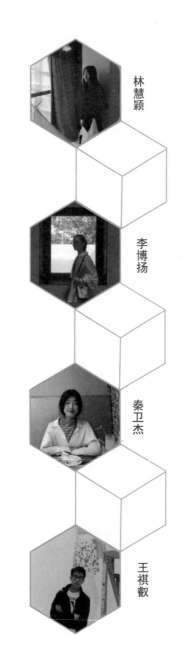

林慧颖

李博扬

秦卫杰

王祺叡

学生感悟
Student's Thought

在这次毕业设计中，我学到了很多新的知识内容，也遇到了一些困难，但是在团队的协作下得到了很好的解决。此外，也要特别感谢指导老师一学期的耐心指导教授，以及组员之间的互相鼓励支持！

作为大学本科最后一场课程作业，任务要求更加得严格，设计也要更加严谨，这次毕业设计让我有了一次更加完整的体验，并最终和小伙伴们一起在欢笑与眼泪参半和日夜不休中顺利完成了项目课题。诚挚感谢老师耐心细致的指导和队友齐心合力的并肩战斗！

通过近一个月的毕业设计，给我最深的感受就是我的设计思维得到了很大的锻炼与提高。作为一名设计人员要设计出有创意且功能齐全的方案，平时就必须留心观察思考，感性认识丰富。感谢老师的指导和督促，也感谢小组成员的通力合作，希望未来还能够一起学习，共同进步。

在设计过程中发现了自己的不足和不少的漏洞，让我能够在以后加以改正，在今后的工作中能够更好地发挥在学校学到的知识，在工作中做得更好。同时感谢老师这一学期的辛苦指导，以及和我一起并肩作战的小伙伴的热心帮助！

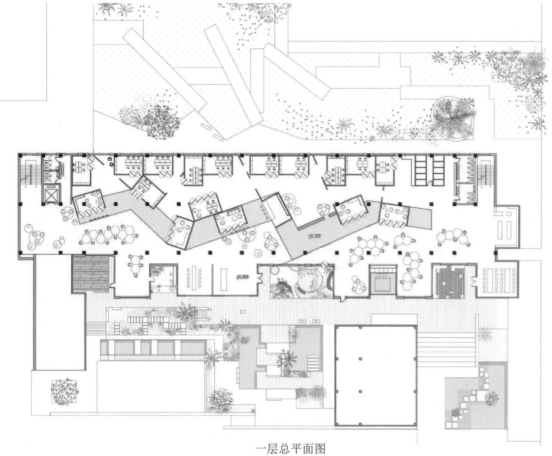

一层总平面图

二层总平面图

一层景观效果图　　　　　　　　　二层景观效果图

雄安设计中心一层以租赁空间为主，设计理念延续原有建筑的微介入手段，保留原有框架，以展览空间为整体空间串联起所有的空间功能，并形成一个完整的展览路线，让租赁公司的企业文化可以展示于此。

展览流线设置为折线形，这种路径比直线更有趣味性，同时扩大了空间中能见的视野范围，尽量减少人群由于展廊路线较长引发的视觉疲劳。

我们还延续了原有多功能模块的想法，设定了固定和自由两种模块。固定模块用于长期办公，自由模块用于非办公区的功能。选用 5.6m×3.6m 的模数。该模块可适用于多种功能形式，比如咖啡厅、会议模式等。

现代的办公空间已经不仅仅只是从业者们的办公区，拥有多种功能并灵活可变的空间才能使空间迸发出巨大的活力。

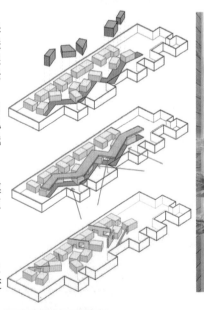

室内庭院

办公模块中可变家具分析

办公模块

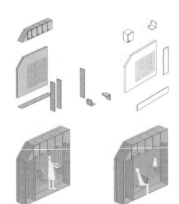

设计师们可以通过带有圆孔的塑料板来动手创造适合自己的家具，让他们拥有创造的乐趣并获得对空间的归属感。

材质采用黄色刨花板，轻薄且采用现场组装的形式。

116

一层天井设计

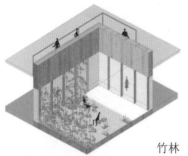
竹林

丽水

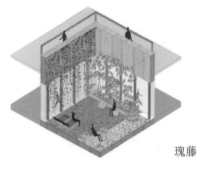
瑰藤

天井设计沿用原有天井的设计概念，每个天井拥有一个独特的主题，这些天井同时又与二层平台有所联系，促使不同高度的人产生连接，促进更多样的行为活动。

天井分为三个大主题，分别为竹林、水园和藤园，它们通过大面积玻璃与共享办公空间相连通，使室内的空间体验升华。

前院景观前后对比

左图是雄安设计中心改造前的景观平面图。为了拉近室内空间与自然的关系，将原有台阶伸长，延伸到门口处。并在台阶上种植植物。大面积的水域为入口处增加活跃感。

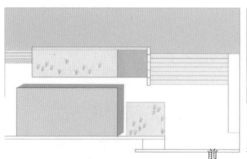
前

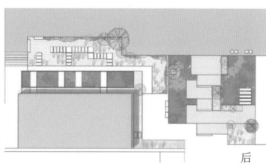
后

答辩评审

原有轴网

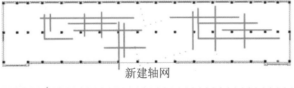

新建轴网

会议室　茶水间

现有平面

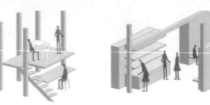

休闲区

种植区

运动区

绿植

亚光大理石
石材地铺
木质地板
青砖地铺
地被层

交通流线

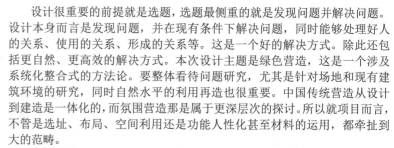

刘恒

设计很重要的前提就是选题，选题最侧重的就是发现问题并解决问题。设计本身而言是发现问题，并在现有条件下解决问题，同时能够处理好人的关系、使用的关系、形成的关系等。这是一个好的解决方式。除此还包括更自然、更高效的解决方式。本次设计主题是绿色营造，这是一个涉及系统化整合式的方法论。要整体看待问题研究，尤其是针对场地和现有建筑环境的研究，同时自然水平的利用再造也很重要。中国传统营造从设计到建造是一体化的，而氛围营造那是属于更深层次的探讨。所以就项目而言，不管是选址、布局、空间利用还是功能人性化甚至材料的运用，都牵扯到大的范畴。

（1）设计从生活和使用作为切入点，围绕着使用来展开是很好的。但对命题一开始的定位不太清晰。命题本身是关于改造，需要对原厂的改造还是改造完之后的改造有一个定位。

（2）原始基调的关联性。要区分与整体整合的部分和具有自己独特的特点的部分，才能明确主题进行深化，避免多方向的浅层的探讨。

（3）手法上的形式比较单一和孤立。

（4）汇报方式要抓住自己的主题，用精练的语言概括特点。

冯任军

（1）该组方案由于想表达的东西有很多，所以表述内容比较分散，没有着重突出方案的亮点和重点。在整个汇报过程中，针对核心内容并不深入。

（2）作为学生来说，尤其是本科生，希望同学们珍惜目前无束缚和制约的教育环境。当真正工作后，来自各方面的束缚就会增多。所以在设计过程中，鼓励同学们发挥天马行空的想象力去拓展自身在设计疆域的知识广度，等进入到实践经验中，再去深入挖掘其深度。在此基础上，整个三维空间体系就会建立起来。

（3）最后一点，作为年轻人（90后或00后），应当思考生命3.0的到来对自己将有什么冲击？作为年轻人，在自身专业领域里需要做什么样的前瞻性来应对时代的巨变？

刘晓军

本组雄安设计中心的方案重点针对年轻设计师们日常工作和生活状态，以及对其所在区域环境的需求，有机地把工作和工作之余的活动空间完美融合在一起，创造出特色的办公环境和氛围。整体设计分析到位、逻辑清晰、设计完成度高。特别把"模块"的运用作为设计切入点，设计了廊架模块、会议模块、公共办公模块、平台模块、茶水间模块等。在方案中对小空间的设计和对模块的组合拼接有创新。另外，把室内和室外框架模块有机地连接，使两个空间巧妙结合在一起，区域之间产生良好关系，这也是设计当中的亮点。

本案重点针对雄安设计中心的公共部分及休闲区域进行设计，建议把设计内容扩展到中心各公司独立办公区域，这样整体的布局和完整性就会更好。

An important premise of the design is subject. The focus of subject is to identify problem and solve problem. The design is inherently to identify problem, solve problem under the existing conditions, and handle the interpersonal relationship, use relationship and formation relationship well. This is a good solution. Moreover, it can contain more natural and more efficient solution. The design theme is green construction, which relates to the methodology of systematic integration. The problem shall be generally considered and researched, especially for the site and existing building environment. The utilization and recycling of natural level is also of great importance. The traditional Chinese construction is integrated process of design and construction, and atmosphere construction represents the discussion in a deeper level. Therefore, in regard to the project, the site selection, layout, spatial utilization, people-oriented functions and material application all relate to a large scope.

(1) The design took the approach of life and use and extends the subject from use, which is a good pattern. However, the initial positioning of the subject is not clear enough. The subject relates to renovation, which requires the positioning as renovation of original factory or renovation after the factory renovation.

(2) The relevance of original tone. The part for general integration and the part with your own unique characteristics shall be distinguished to specify and deepen the theme and avoid the shallow discussion in various aspects.

(3) The forms of the techniques are relatively unitary and isolated.

(4) The report shall focus on the theme and summarize the features with precise language.

(1) Since the scheme of the team plans to express many things, the representation is relatively incompact and does not emphasize the scheme highlights and key points. The overall report does not thoroughly explain the core contents.

(2) We hope that the students, especially undergraduates, can cherish the unrestricted and unlimited education environment. When you start work, there will be more and more restrictions from various aspects. In the design process, we encourage the students to give play to the unconstrained imagination, expand the knowledge scope in design field, and thoroughly explore the depth in practice experience. On this basis, the three-dimensional spatial system will be established.

(3) At last, the young people (post-90s or post-00s generation) shall think how the arrival of life 3.0 will influence themselves. How shall the young generation deal with great times change in their own fields with forward-looking perspective?

The scheme of Xiong'an design center of this team focuses on the daily work and life status of the young designers, and the demands in the respective area environment, organically integrates the activity space for work and after work, and creates featured office environment and atmosphere. The overall design contains accurate analysis, clear logical representation and well completed design contents. It especially takes the design approach of "module" application, and designs gallery frame module, meeting module, public office module, platform module, and pantry module, etc. The scheme innovates in the small space design, and combination and connection of the modules. Moreover, the indoor and outdoor framework modules are organically connected to subtly combine two spaces, and thus develop good relation between the spaces. This is a highlight.

The case focuses on the design of public area and leisure area of Xiong'an design center. It is recommended to expand the design contents to the independent office area of the companies to improve the general layout and completeness.

建设可再生未来——雄安设计中心室内外环境设计

Building A Renewable Future—Interior and Exterior Environment Design of Xiong'an Design Center

高　　校：西安建筑科技大学

College: Xi'an University of Architecture &Technology

学　　生：邓莹、麦世星、于硕、秦致远

Students: Deng Ying, Mai Shixing, Yu Shuo, Qin Zhiyuan

指导教师：刘晓军

Instructors: Liu Xiaojun

课题分数：85

Subject Scores: 85

邓莹

麦世星

于硕

秦致远

学生感悟
Student's Thought

作为本次竞赛小组组长的我，一开始真的是压力重重，面对竞赛出现的很多问题，我们具体问题具体分析，当遇到很多我们难以解决的问题时，我们小组积极主动寻求老师以及很多优秀学者的帮助，最后突破解决了很多难题，历时四月，毕业设计圆满结束，感谢老师的悉心指导，感谢组员的辛苦付出。结束既开始，愿我们在未来的路上不负韶华，砥砺前行。

作为大学本科最后一场课程作业，我感到很荣幸能有这个机会参加本次比赛，本次课题从设计实际问题层面的理念探讨到室内装配式空间的未来发展，在全方位多层面上均有探究，让我们学到了很多在学校没有接触到的实际设计问题。从理念分享到方案成形，通过和有关专家、学者们一起探究，最终和小伙伴们一起，在欢笑参伴和日夜不休中顺利完成了项目课题。最后诚挚感谢老师耐心细致的指导和组员齐心合力的努力！

通过一个学期的竞赛学习，我对室内设计有了更深入的了解，从参加这个竞赛开始，每一天都过得很充实。小组的精心准备，全力以赴，虽然没有得到最好的名次，但是我们收获了友情和经验，丰富了设计经历和知识的阅历。让我们在整个完成竞赛的过程中学习到更多实际专业知识。名次不重要，重要的是过程。

通过本次比赛的历练，以室内装配式空间为契机，让我在设计方面有更加全面的理解和探究，也认识到自己需要改善进步的地方，吸取很多专业上的知识，学习了在学校里学不到的东西，真的让我开拓了视野、拓宽了设计领域的格局，同时也很荣幸认识了很多志同道合的朋友和精英，希望未来还能一起学习，共同进步。

概念提出 | Proposed Concept

魔方是一种由不同体块组成的创意玩具，在空间上具有很强的可变性。如今已不仅仅局限于六面魔方，更多的高阶魔方、创意魔方更是多种多样，组成魔方的体块也有了很大的变化。在魔方的转动、拆卸过程中，可以发现组成魔方的各种体块，在空间上能创造出无限的可能性。

我们利用了魔方的特性，通过模块，转化成不同的元素，支撑构件、墙板、家具，来进行各种不同的组合，形成了几种不同的空间形式，来满足办公需求，并将魔方元素衍射到其他室内空间以及室外环境，丰富室内外空间，满足人群使用需求，然后植入绿色理念，让整个魔方元素富有更深层次的内涵。

在探讨了多种魔方形式后，我们对五魔方进行了变形处理，基面为六边形，使得整个体块变得更便于拼接，也可以更好地融入场地，在进行切割变形之后，生成了不同的体块，根据摆放位置，可以转化成不同形态的构架、隔板、家具，但都可以进行自由拼接组合，形成或开敞，或围合或半围合的空间，以适应个人的办公习惯，用体块生成的家具，也可以根据分工类型，个人习惯，进行不同的变形，来更好地适应各自的工作，提高工作效率。

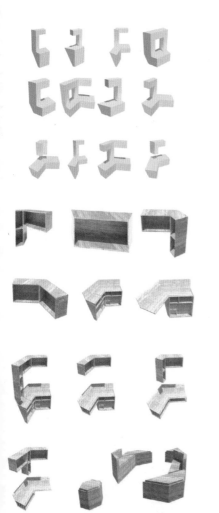

家具衍生

体块拆分

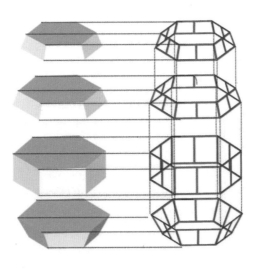

构架生成

123

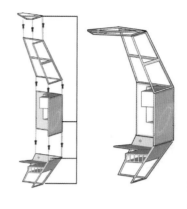

对魔方进行点、线、面不同层次的提取，转化为魔方工位的不同构件进行组合。

用户根据需要选择工位系统的组成部分，形成工位系统基本单元，通过对基本单元的不同组合，形成不同的办公空间，适应多种办公需要。

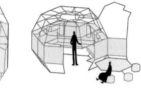

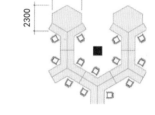

2600

2300

基本单元

两组形成半围合空间

多组形成围合空间，局部形成讨论区

多样化组合，组成多种办公空间

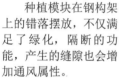

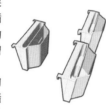

种植模块在钢构架上的错落摆放，不仅满足了绿化，隔断的功能，产生的缝隙也会增加通风属性。

通过塑料小模块的种植构件，组成绿化墙板，工厂预制模块，灵活性强，可自由拼装，施工工期短。

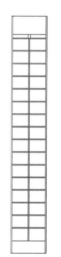

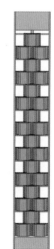

124

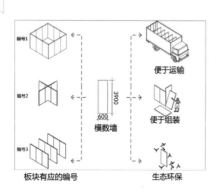

编号1

编号2

编号3

便于运输

3900

600
模数墙

便于组装

板块有应的编号

生态环保

就近选择厂家，进行模块编号，在现场利用 APP 等手段，监控每一个模块的安装情况，并及时调整。

大楼建成后，利用 APP 等手段，对大楼的节能属性进行监督，促进办公人员的各种绿色行为。

运用智能监控，实时收集大楼的能源收集情况，包括电能、太阳能、生物能等，反馈大楼内的能源损耗，达到节约能源的作用，通过对装配式构件的不同组装方式，解决大楼内的通风采光等要求。

秸秆板　连接构件　木质龙骨

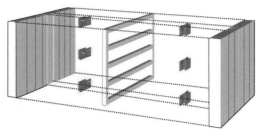

玻璃隔墙的设计主要为了满足室内的自然采光，减少室内的能耗，秸秆板使用当地原生材料，秸秆再次利用，减少一定的排放。

不同材质的墙板进行不同的组合，形成不同的空间，满足不同的办公空间需求。

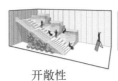
开敞性

半开敞性

封闭性

综合灵活性

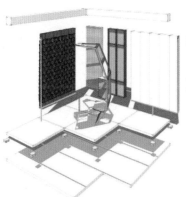

地板进行架空处理，所有管线走地板下既减少了楼层与楼层间的噪声，也方便安装与后期维修。

胶囊房为办公人员短时间休息设计，可容纳一些小范围的安静活动，办公人员可以在这里短暂午睡，也可以在这里冥想、阅读等。

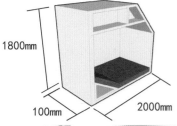

1800mm
100mm
2000mm

一层接待区效果图

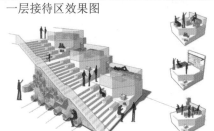

小型会议，休息

展览图纸

头脑风暴时刻

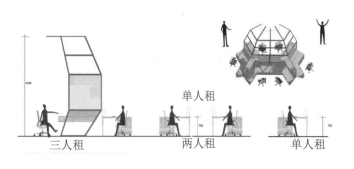

单人租

三人租　　　　两人租　　　　单人租

楼梯作为一层和二层的主要交通空间，采用了灵活多变的设计手法，运用魔方元素，进行切割演变，形成可变性的交通空间，楼梯分为半围合的小空间组成，根据参与者自由变换选择空间，楼梯的墙面运用魔方转变演化成装饰的六边形框架，可做小书柜，也可做成桌椅。

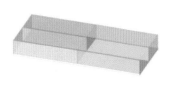

二层为四家小型公司，中部形成交流、休憩展示空间，制作装配式隔断，可根据需求形成不同的空间形式，方便办公活动的展开。

场景一

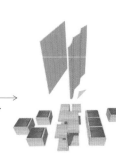

600mm×3900mm×40mm 可折叠板

450mm×600mm×600mm 可移动座椅

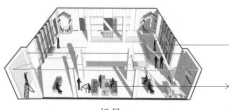

场景二

展览挂图区域

公共交流区域

二层的公共空间是大楼的最中心点区域，设定空间是灵活多变的，可以形成不同的空间需求。

三层综合阅读室效果图

三层为两家大型公司，两个公司中间设置公用的大型会议室，节约空间，体现了共享的理念。另外设置一些休憩区、胶囊房，体现人文关怀。

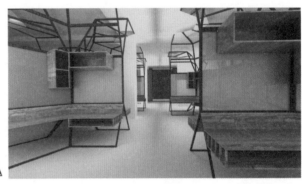

四层节点效果图

四层为三家中型设计公司，每家公司功能完善，但面积会较为不足，所以将会议室、洽谈室、胶囊房进行共享，节省出空间，在相互独立的基础上，也能产生联系。

五层工位布置节点

五层为一家大型公司，在功能上，具有自己独立的办公区、会议区、休闲区、洽谈区，也能分割出不同规模的办公区，以适应不同的办公需求，因为处于最高层，单独设置了打印间与休闲娱乐室，为员工提供方便。

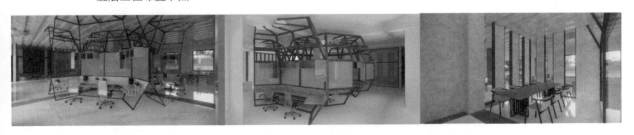

五层节点效果图

景观展示 | Landscape Display

景观模块分析

拆分

重组

通过魔方的拆分、重组，生成景观当中的设计元素，特定的模块都可以通过工厂预制，形成装配式的模块，创造以魔方为主题元素的设计结构，创造外环境的魔方空间。

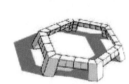

模块的多种可能性 = 景观小模块 + 景观空间模块

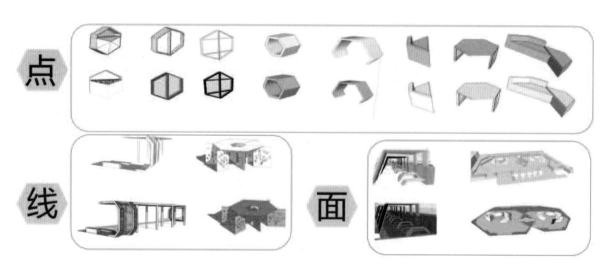

点

线

面

问题发现和解决策略 | Problem Discovery and Resolution Strategies

问题

绿化空间形式单一

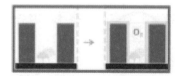

解决策略

增加垂直绿化和屋顶

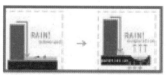

空间缺乏体验

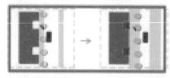

设置可移动的流动亭

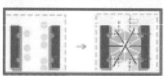

空间缺乏安全感

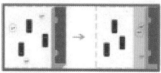

释放道路部分，增加更多的休憩场所。

辅助空间缺少

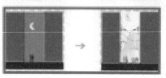

大楼内禁止抽烟，增加多功能的休息室，外卖取餐处，快递收发室，增加人文关怀。

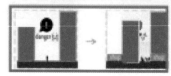

「室内设计 6+」2019（第七届）联合毕业设计
"Interior Design 6+"2019(Seventh Year) Joint Graduation Project Event

屋顶花园平面

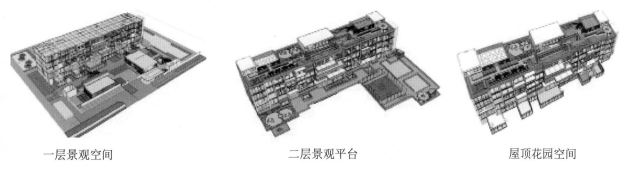

一层景观空间　　　　　　　二层景观平台　　　　　　　屋顶花园空间

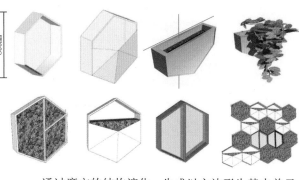

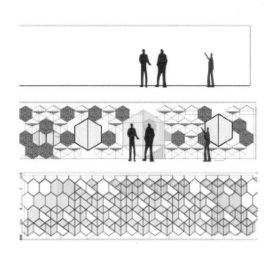

　　通过魔方的结构演化，生成以六边形为基本单元的花池、玻璃框架、种植槽，可以组合成独具特色的绿植墙，以及六边形组合的半透明的玻璃隔墙，突显出以六边形为基本的元素，创造装配式设计和绿色营造的完美结合。

地面结构解析 ┃ Ground Structure Analysis

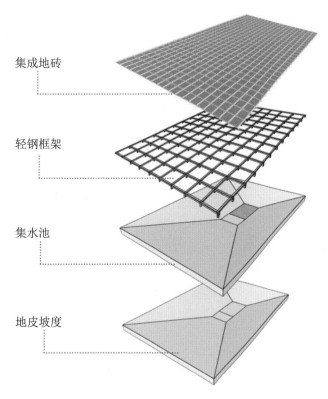

　　集成的地砖减少现场的湿作业，减少现场的施工时间。

集成地砖

轻钢框架

集水池

　　景观的中间处设置多种类型的集水池，用来收集整个大楼的雨水，便于大楼内的水的利用和排污。

地皮坡度

一层景观

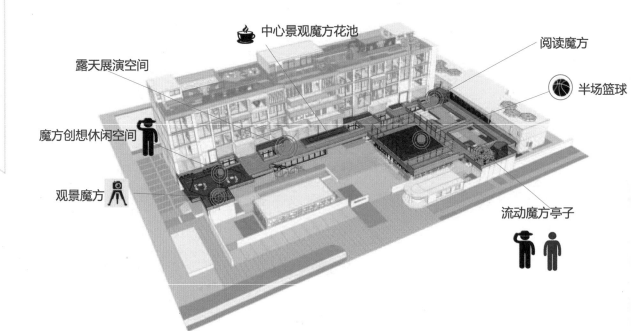

吸烟室

会议大厅

露天展演空间 👷

☕ 魔方亲水休闲区

魔方快递驿站

吸烟室

SF EXPRESS 顺丰速运

运动健身空间 🏓🏸

用餐处 🍴

入口空间 🧑‍💼

屋顶花园平面

鸟瞰图

☕ 中心景观魔方花池

阅读魔方

露天展演空间

🏀 半场篮球

魔方创想休闲空间 🧑‍🌾

观景魔方 👷

流动魔方亭子 🧑‍🌾🧑

屋顶花园效果图

空间人物串联

景观整体的功能设计充分考虑了人性化的需求，创造了以人为本的现代景观。

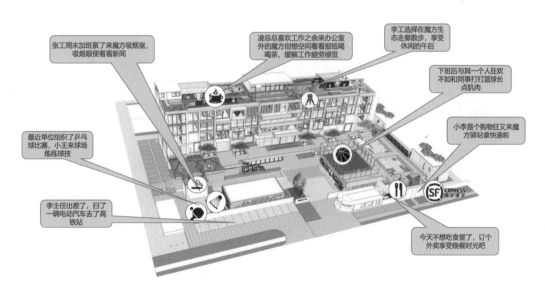

空间剖面图

景观环境和室内联系通过屋顶花园的雨水收集，可以给室内提供绿植墙的滴灌提供充足的水源。

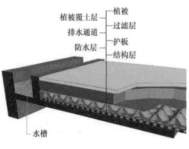

王的刚

我们希望模块化的绿色未来能形成更广泛的理念。魔方是一个非常好的六面立面体。

（1）方案阐述内容稍有模块化，最好汲取新出现的理念。

（2）团队合作很重要，本次毕业设计是依靠合作完成的，一定要合作，才会给人惊喜。

（3）选题很讲究，方案明确了 12 个企业员工对于场所的使用需求，包括员工急需解决的问题和如何解决，比如说员工的压力问题。

（4）分析方法很重要。首先，在改造优秀的建筑时，前期工作和设计方案这两个环节都是需要去慎重分析思考的。针对现状分析，深入挖掘整理好的内容和方面保留下来，去其糟粕、取其精华。面对不好的部分要及时修正过来。其次，在更改完增加的内容基础上回过头来验证一下方案是否解决了问题。最后，把方案做出来达到好的程度上，也要注重汇报时把逻辑关系串清楚。

左琰

整个汇报思路清晰，分析较完整。该设计题目包含 3 个方面：

（1）老厂房改造。

（2）办公室室内设计。

（3）装配化主题。整个方案推出了"魔方"理念，运用魔方的单元组合来布局展开。设计后的空间较现状丰富了，功能也得到细化。这个方案对于设计竞赛倒有新意，但实际操作会缺乏一些落地性。由于原有办公室已完成内饰，且是低成本的绿色营造，因此新推出的家具模块化属于家具空间一体化，从家具层面上来说优势明显，如提高收纳空间、视线避免干扰、空间更人性化，但也势必会追加成本，与装配化设计的初衷背离。此外"魔方"的形态变化比较受限，且形成的局部空间因封闭而使通风和采暖会受阻碍，易造成忽冷忽热的现象。

132

王婷婷

当前 90 后或 95 后的年轻一代设计师对舒适感非常重视，尤其是在室内设计中。大学时在设计院实习，最常见的就是一地的折叠床和沙发垫等。所以方案中设计的午睡胶囊在设计院里推行应该是非常受欢迎的。在室内环境的舒适性上，尤其是学环艺专业的同学，可以考虑一些环境心理学的问题。在材质的选择上，比如说质感、触感等，这些设计要素在本次的方案中没有涉及，还有色彩方面的视觉感受也未提及。而该方案以魔方为元素，魔方最大的特点就是不同颜色的面体，因此色彩会有很多种，但在方案的最终效果图里并未表现出来丰富的色彩。可以在后续调整中考虑在不同的空间里采用更有意思的、多样化的色彩，同时更符合年轻人的审美。

We hope that the modular green can develop into a more extensive idea in the future. Rubik's cube is a good hexahedron.

(1) The scheme explanation is modularized to certain degree. It is suggested learning from the new ideas.

(2) Teamwork is very important. The graduation design shall be completed through cooperation. Cooperation will bring surprise.

(3) The subject is well selected. The scheme specifies the demands of 12 enterprise employees in the site, including employees' problems to be urgently solved, the way to solve, and employee pressure issue.

(4) Analysis method is also important. Firstly, we need to carefully analyze and think about the two procedures of preliminary work and design scheme when renovating excellent buildings. After status quo analysis, the good contents and aspects shall be thoroughly explored and sorted to remove the dross and absorb the essence. The poor part shall be revised timely. Secondly, we shall verify whether the scheme really solves the problem after adding new contents. At last, the scheme shall be well prepared and attention shall also be paid to clearly sort out the logical relationship.

The overall report is given with clear thinking and relatively complete analysis. The design subject contains 3 aspects:

(1) Old factory renovation.

(2) Interior design of office.

(3) Assembly subject. The scheme releases the concept of "Rubik's cube", and develops the layout through the unit assembly of Rubik's cube. With the design, the space is more diversified than the current status, and the functions are refined. The scheme is creative in a design competition, but lacks feasibility in the actual operation to certain extent. Because the interior decoration of the office has been completed in the form of green construction at low cost, the new furniture modularization belongs to furniture and space integration. From the perspective of furniture, this is a great advantage to expand storage space, avoid sight line interruption, and make the space more people-oriented. But it also increases the costs and conflicts with the original intention of assembly design. Moreover, the form changes of "Rubik's cube" is relatively restricted, and the local areas thus developed will face ventilation and heating restrictions and tend to cause the sudden changes of temperature due to the closed state.

The young designers from the post-90s or post-95s generations currently attach much importance to comfortableness, especially in interior design. When I took the internship job at the design institute when I studied at college, I always saw the folding beds and sofa cushion, etc. therefore, the afternoon nap capsule in the scheme design shall be very popular if we publicize in the design institute. In regard to comfortableness of indoor environment, especially the students in environmental art design major can consider the issues in environmental psychology. As for material selection, the design factors including texture and sense of touch are not covered in the scheme. The visual perception for colors is also omitted. Since the scheme adopts the element of "Rubik's cube" and as we know the most significant feature of Rubik's cube is the surfaces in different colors, there shall be many colors. The final rendering of the scheme does not represent rich colors. It is suggested using more interesting and diversified colors in different spaces in the subsequent adjustment, which will better match the taste of the young generation.

133

高　　校：北京建筑大学

College: Beijing University Of Civil Engineering And Architecture

学　　生：崔雨晨、赵佳慧、柴鑫

Students: Cui Yuchen, Zhao Jiahui, Chai Xin

指导教师：杨琳

Instructors: Yang Lin

课题分数：90

Subject Scores: 90

崔雨晨

赵佳慧

柴鑫

学生感悟

Student's Thought

　　我非常高兴能够参加这次"室内设计 6+"联合毕设，通过这次毕设，我成长了很多，也收获了很多，感谢我的老师和同学们，他们给予我很多帮助。设计之路还很长，我会继续努力，坚持自己所长，弥补自己不足，让自己变得越来越优秀。

　　此次课程中途不免有很多困难，比如改图、否定方案这些家常便饭，但我们都彼此陪伴着坚持了下来，半年的努力就此告一段落了，大学生活也就此结束了。等待我们的或许是艰难险阻又或许是一片光明，无论何去何从，都请坚持下去。衷心感谢各位指导老师的耐心指导与谆谆教诲。

　　如果生活中有什么使你感到快乐，那就去做吧，不管别人说什么，Be happy。

展陈设计 | Display Design

　　整体空间分为四部分，分别为展示区、放映区、卫生间和公共区域。按照交通流线来布置展陈形式，从入口开始分别为序厅、零碳历史沿革、零碳材料展示区、零碳案例展示区、零碳放映区、零碳实物展示与卫生间零碳体验。参观者可从历史到案例再到实际体验来连贯性的熟悉并了解且应用零碳技术，达到了解和宣传的双重目的。

　　整体设计主旨采用装配式的主题，通过装配式的六面和展陈形式来展示零碳技术。

功能分区 | Function Division

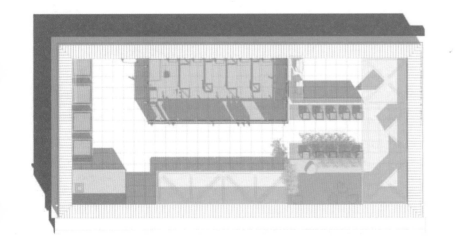

- 展柜展示
- 序厅
- 绿化
- 放映区
- 展架展示
- 卫生间
- 交通

135

装配式设计 ┃ Assembly Design

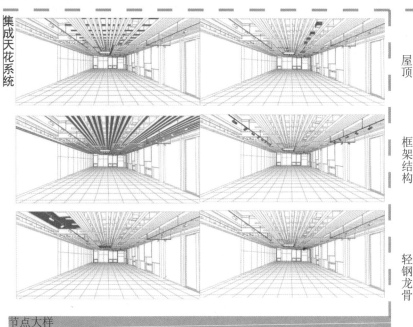

集成天花系统

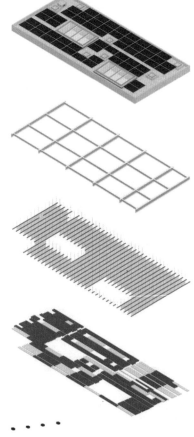

屋顶

框架结构

轻钢龙骨

铝方通吊顶

照明系统

节点大样

天花采用集成天花吊顶系统，用阳极氧化铝材质制成的铝方通，尺寸为 2100mm、2500mm 和 800mm 三种，分为木色和白色两种颜色。

照明系统分为重点照明和均质照明两种，对于展柜和展架采用重点照明，交通和放映区为均质照明，采用滑轨射灯、吸顶灯两种，吸顶灯尺寸为 650mm×650mm 和 250mm×250mm 两种。

集成墙面系统

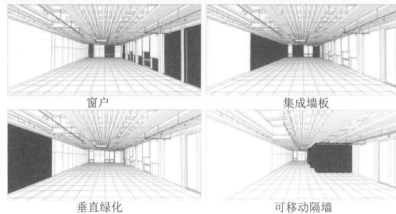

窗户 　　　　　　　集成墙板

垂直绿化 　　　　　可移动隔墙

节点设计

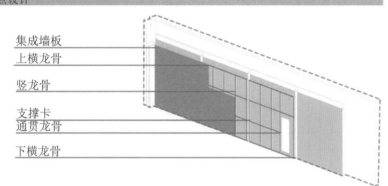

集成墙板
上横龙骨
竖龙骨
支撑卡
通贯龙骨
下横龙骨

可移动隔墙

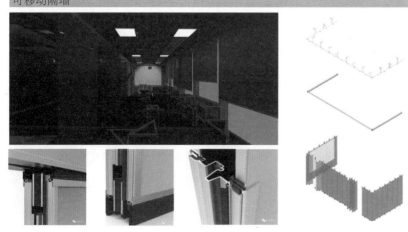

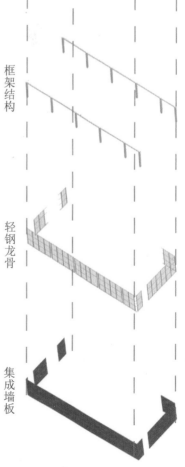

木格栅外墙

混凝土墙体

装配墙体

框架结构

轻钢龙骨

集成墙板

137

集成地面系统

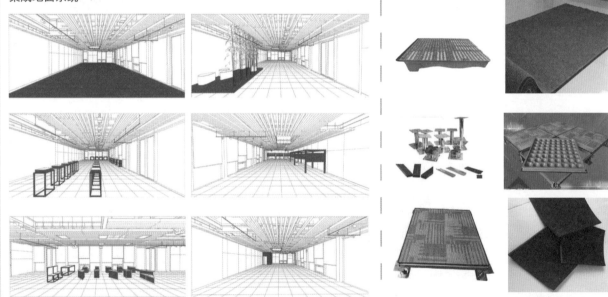

节点设计

1. 清洁地面

2. 画网格线

3. 放置支架

4. 调整水平

5. 横梁联接

6. 安装地板

7. 四周封边

8 清洁表面

地面采用集成地板系统，使用 600mm×600mm 的地毯和碳酸钙板结合，通过 100mm 高的龙骨支撑，搭建整个地面系统。整体颜色分为 3 种，分别为黑、深灰、浅灰，通过交叉拼接组合，达到不同的效果。

地板采用泰腾地板，地毯纹架空地毯全钢陶瓷面防静电地板，很好地满足了装配式的主题。

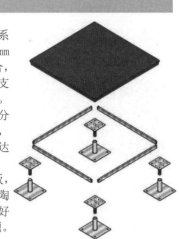

装配式展陈系统

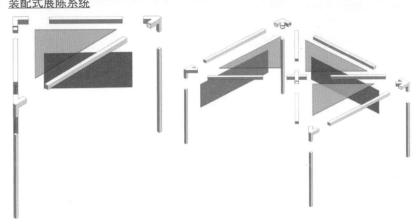

展架组合

展架采用木棍和连接件连接组合而成，棍的尺寸分为3种，分别为2100mm、1800mm和2500mm。通过3种连接件可以组合成不同的形状，满足不同的展陈需求。展示内容采用磨砂纸或织物通过打印的形式进行展示，采用胶粘的形式，可拆卸并可多次展示。

连接件

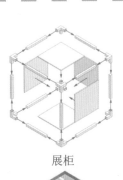

展柜

桌椅

桌子

绿植

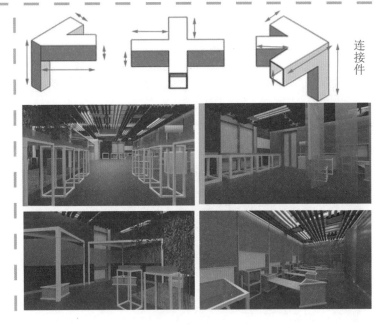

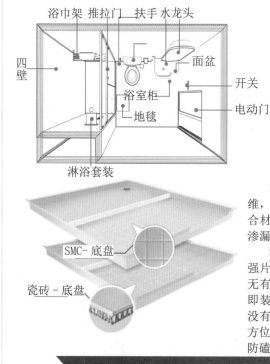

浴巾架 推拉门 扶手 水龙头

四壁

面盆

浴室柜

地毯

开关

电动门

淋浴套装

SMC 顶板

HCM 壁板

HCM 壁板

SMC 防水底盘

SMC- 底盘

瓷砖 - 底盘

SMC 原料是铝蜂窝芯、聚合剂和玻璃纤维，通过工业高压一体成型，铝蜂窝芯符合材料零吸水特性，从根本上根治卫生间渗漏和异味。

SMC 复合材料是不饱和聚酯玻璃纤维增强片状模塑料，是经高温一次模压成型，无有害气体，无放射性元素，绿色环保，即装即用。SMC 材料分子结构紧密，表面没有微孔，不藏污垢，圆弧边角设计，全方位无死角，卫生清洁。质感湿润，不坚硬，防磕碰、保温，更有按摩双脚的作用。

6 hour

传统弊端

墙面系统

HCM 丽晶
复合装饰板

SMC 航空树
脂复合材料

工作模型 | Work Model

不同颜色表示不同的装配模块

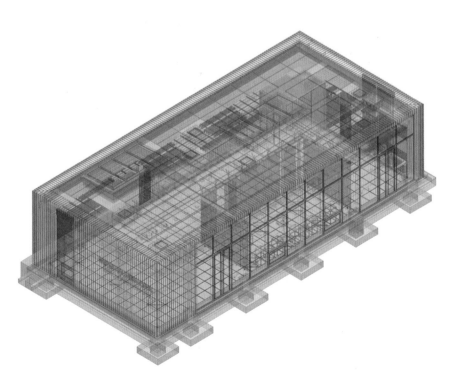

数字化模型

吕勤智

该组方案的成果很完整，表现力比较强。针对绿色设计和装配化主题提出了有创新性的设计方案。同学们抓住了绿色营造的主题理念，从装配化手段的角度来切入，从技术分析入手到问题的解决，较好地提出合理的设计方案。方案需要进一步完善零碳展示馆的空间流线交通内容。参观者从入口到出口，空间的交通流线需要做得清楚明确，才有益于引导观众的观展顺序，满足人在空间当中的感受和体验；整个展示厅的面积是有限的，在这个小空间里面，要思考如何有效利用空间，增强观众的空间体验感。除此之外，还应该在空间中更多运用垂吊的方法丰富展示效果，更有效利用有限的空间。针对小空间多从设计手法的角度来丰富空间设计，这方面还可以进一步推敲。

蒋浩

该组方案设计的定位是展示空间的设计，众所周知其是室内空间设计的其中一部分。首先要明白室内空间的设计内容。室内设计应该包括空间、材质、色彩、软装陈设和环境心理的设计，而这里过多地展示了技术层面上的东西。尽管主题是绿色营造，但展示空间的设计还是需要整体来加以说明。方案在汇报过程中的完整性需要再增强一些。另外，这是毕业设计答辩的形式，选择毕设题目的原因和呈现的内容以及最终要达到什么样的空间效果，这在毕业设计过程中要体现出来。而且通过毕业设计，反映出对这个时代和社会的应用意义，尤其是对人文情怀方面的关注。再者也需要增加一点环境心理学方面的内容，应该考虑到来自不同区域的人们参观时的感受等，总而言之要体现出以人为本的设计理念。

142

曹阳

毕业设计要完成四个阶段的内容才能算好的毕业设计或作品：主题性、方法性、展示性、创意性。

（1）从主题性来讲，今年"室内设计6+"的主题是绿色营造，室内装配化才是真正的方法。该组这点做得非常好，包括修改之后的方案，实用性很高。

（2）该组同学掌握了装配式技术的方法，但是相对来说，只是达到了认知和对现有产品利用的阶段。

（3）在创意不足的前提下，应该先了解该事物，以现有的事物来丰富你对室内设计装配化的利用方式。

（4）展示性略差，技术展示太多。一定要在一个空间、一个角度上做出多种空间变化的可能性。

（5）在创意性方面待提高，但是从开题到中期再到现在，同学们处于不断成长的阶段。

The scheme of the team represents complete result and relatively strong expressive force. On the subject of green design and assembly, the team proposes innovative design scheme. The students grasp the theme concept of green construction, take the approach of assembly means, and suggest relatively reasonable design scheme in the process from technical analysis to problem solution. The scheme shall be further improved in regard to spatial traffic flow contents of zero-carbon exhibition hall. The space traffic flow of the visitors from the entrance to the exit shall be clearly designed to help guide the visitors to follow the exhibition view sequence and satisfy the people's feelings and experience; the exhibition hall covers a limited area. In this small space, we need to think how to make use of the space effectively and enhance the spatial experience of the audience. Moreover, we need to demonstrate the performance in the method of hanging in a diversified manner in the space, and make more effective use of the limited space. More considerations shall be made to diversify the spatial design in small space from the perspective of design technique.

The scheme design of this group is positioned as design of exhibition space, which is known to be a part of interior space design. First of all, we must understand the design content of the interior space. Interior design shall include space, material, color, soft furnishings, and environmental psychology, but there are too many technical aspects here. Although the theme is green construction, but the design of display space still needs to be explained as a whole. The integrity of the solution in the reporting process needs to be enhanced a bit more. In addition, this is the form of graduation project defense, the reason for choosing the graduation project title, the content presented and what kind of spatial effect to achieve, which shall be reflected in the graduation project process. Moreover, the graduation design reflects the application significance of this era and society, especially the concern of humanistic feelings. In addition, some content of environmental psychology shall be added, which shall take into account the feelings of people from different areas when visiting. In a word, it shall reflect the people-oriented design concept.

Graduation design can only be considered as a good graduation design or work after it meets four features: Theme , Methodology , Demonstration , Creativity.

(1) From the perspective of theme, the theme of interior design 6+ this year is green construction, and the indoor assembly is the real method. The group performed extremely well for this aspect, and the revised scheme is with high practicality.

(2) Students in this group have mastered the methods of assembly technology, but relatively speaking, they only reached the stage of cognition and utilization of existing products.

(3) Under the premise of lack of creativity, you shall first understand this thing and enrich your use of interior design assembly with existing things.

(4) The display is slightly poor, and the technology shows too much. Be sure to make the possibility of many spatial changes in one space and one angle.

(5) Creativity needs to be improved, but students are in the stage of continuous growth from the proposal to the middle stage and up to now.

答辩评审

高　　校：南京艺术学院

College: Nanjing University of the Arts

学　　生：陈涛、李艺蓓、叶子萱、李赟

Students: Chen Tao, Li Yibei, Ye Zixuan, Li Yun

指导教师：朱飞

Instructors: Zhu Fei

课题分数：85

Subject Scores: 85

中国国际进口博览会展示设计

Exhibition Design of China International Import Expo

学生感悟
Student's Thought

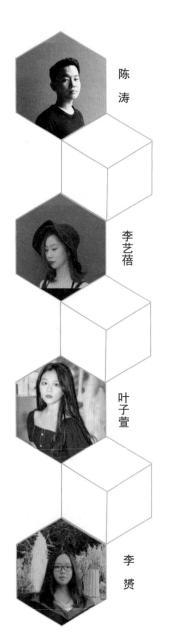

陈涛

感谢我们的指导老师朱飞教授对我们的指导与帮助，同时感谢主办方搭建的这个平台，给了我们一次展示的机会。另外在设计过程中，在朱飞教授的指导下，我们通过合理的分工，紧密的合作，有条不紊的有序推进，活动的过程本身就是一场学习。

李艺蓓

这次联合毕设，让我总结出四个字——越挫越勇，不要被自我感觉所蒙蔽，还有许多要学习的地方，要多多努力，尽量多的抓住机会，提高自我的表现潜力，从每一件事中找到进步的目标，让自我变得越来越优秀。

叶子萱

从开题到中期到最后答辩的这几个月时间里，我经历了很多，从最初满怀希望的开始到准备中的挫折，在这个过程中我不断的体会着每一个细枝末节，每一次老师给我指出错误我都要去反复的琢磨，反复的改，遇到了很大的麻烦，幸好最后都一一解决。

李赟

参与此次"6+"活动，与各大高校同学、老师及行业中的优秀前辈们交流学习，是我整个本科学习中非常宝贵的经历。让我不仅对绿色搭建有了认识和了解，也对设计有了新的感悟。再次也要感谢我的毕设指导老师朱飞老师对我的悉心指导和支持。

项目定位 ┃ Project Orientation

营造独特性格

建立可识别性

提高可达性 + 可视性

价值倍增布局

绿色装配化搭建

展厅出入口要方便进出，内部流线要合理，提高空间容纳量
展厅功能设计上要不同身份观众的功能需求
展示内容从物质转向非物质

展示手段从单向转向双向互动
展示形式从长久不变转向临时多变

灵感来源 ┃ Inspiration Source

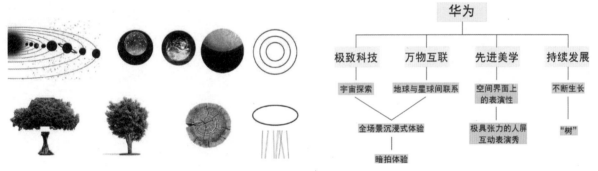

本作品从宇宙星际衍生出空间主体形态，以宇宙遨游为流线规划，以万物互联为主要线索，以树的生长为动态线索，体现科技飞速发展仍不断探索的企业精神。整个空间中的色彩多以白色与科技灰为主，以便接收投影成像，投影多选用宇宙星体的变幻图案。

平面规划 ┃ Plane Planning

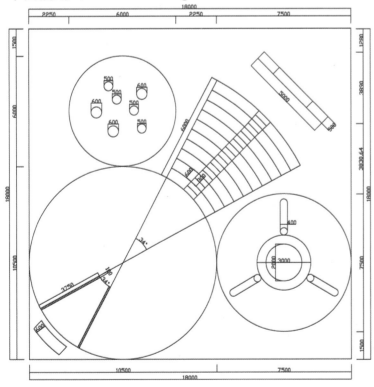

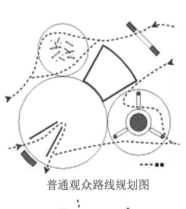

普通观众路线规划图

专业人士路线规划图

145

展区分为四大板块，A 区域（A1 科技技术展区、A2 高台）为整个空间中的主体，以 45°轴线向两边分散出 B 区域（展品展区）和 C 区域（企业形象展区），并在轴线对应的另一端设置 D 区域（未来愿景展区）。

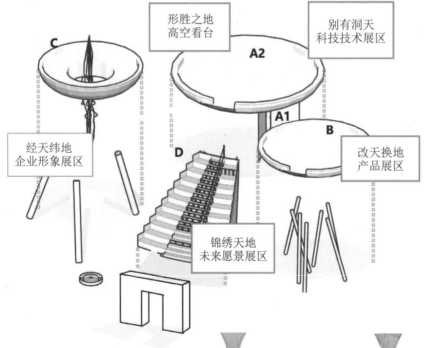

形胜之地
高空看台

别有洞天
科技技术展区

经天纬地
企业形象展区

改天换地
产品展区

锦绣天地
未来愿景展区

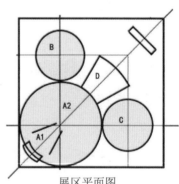

展区平面图

A1 区域正立面图

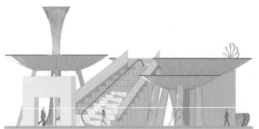

D 区域侧立面图

内容设计 | Content Design

　　A1区域是名为"别有洞天"的科技隧道，由拼接成墙的手机组成。手机幕墙的画面由终端设备控制，其上图案可随观众行走而变化，手机幕墙的另一面则是有许多均匀小孔的透光墙，观众可将透明亚克力棒随意嵌入小孔中，感受到华为企业为观众需求而不断改变的企业原则。

　　A2区域名为"形胜之地"，是将整个空间中最主要的高层看台利用起来，可登高的设计象征着华为企业不断攀登，在新的领域占据形胜之地的企业目标，而看台的底部弧面则是星体的投影，呼应宇宙遨游主题也强调万物互联的目标。

　　B区域名为"改天换地"，是实体产品展区，上层是与A2区看台形状一致但不支持观众登高行走的弧面结构，下层是由真实的支撑结构柱和一些装饰性结构柱组成，前者固定上层，后者开洞作为产品的展台，放置华为的最新科技产品，如P30PRO、MATE20PRO、VR2等。柱身投影出产品名字，柱底是不断变换的地面投影装置。

　　C区域名为"经天纬地"，是华为的企业形象展区，是一个用丝线和灯光来打造的装置，以此呼应主要线索：万物互联。B区域与C区域的形态都来源于树的生长，取聚树成林的寓意，强调华为企业文化中的凝聚力和不断进行科研、积累经验、完善自身的决心。

D区域主要由阶梯开敞看台、可登高的楼梯以及模块化结构组成且设置有定点互动装置的梦想之门组成。D区域是企业未来愿景区，在这里，节节高的看台代表着用户的监督与参与对企业的重要性，可以穿过的梦想之门代表一道寻求突破的屏障和不断提升的目标，代表跨出或跨入这个门都将是一片新天地。来往的观众将通过定点互动装置在变成别人眼中的表演者，将用户与企业紧密关联，在这个区域打造出一个"观众即演员"的舞台，除此之外也会安排不定时的机器人舞蹈以增强空间的表演性。

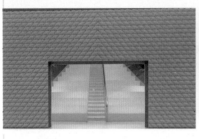

吊顶示意图　　　　　　　　　　　　　　配件示意图

绿色办展是国际进口博览会高品质展会的一项标准，进口博览局制定了《绿色中国国际进口博览会标准》，结合本次"6+"联合毕设活动"绿色营造——建筑室内装配化设计"的主题，我们本次毕设的企业馆设计中绝大部分选用了装配化搭建，在降低展会成本的同时，也增强空间的规律。

凭借霍克公司技术部的帮助，我们将电脑中虚拟的模型演变成可落实的设计，利用霍克公司的材料还原设计中的每个结构，遵从轻松布展、不浪费材料、绿色环保的搭建原则，完成了我们的华为企业展位。

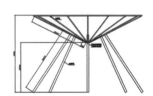

形胜之地搭建结构图

模型展示

模型展示

王的刚

（1）命题是至关重要的，学生解题思路首先是从命题上去推导。

（2）设计工作可以从功能层面深入分析课题内容。

（3）以绿色搭建为前提，设计成果不仅要在技术层面上有表述，还要建立更广泛的概念。其不仅仅只是技术层面的问题，它也是属于理念的问题。这个理念会对设计逻辑结构产生什么样的影响？这个问题对学生来说重要的是首先建立一个体系。

判断生命脉动的立意，产生类似胶囊的形体，从生命脉动推到胶囊，这个形式最终产生的结果并不是最重要的。相反，生命脉动如何推理到胶囊的中间过程才是最重要的，否则这个推理能推出很多不同结果。依据答辩叙述来看，立意和结论之间，生命的脉动是否显示是一阴一阳？设计结论中涉及的概念很多，例如太空舱等。再谈形式表达，生命脉动的立意是如何落到形式上？再把形式和布展内容的关系建立，形成清晰的逻辑关系。比如说以何置于阳、以何置于阴。总而言之同学们阐述的过程是不错的，把宇宙和万物互联起来，通过科技分析进而组合该空间，这一方法形式是很好的。关于该设计，应该思考当形式完成后，空间该是如何产生的？围合产生的活动场景空间又是怎么和构筑物划分界定的？

王祖君

展示设计在室内设计范畴中的复杂程度是非常高的，需要清楚确定展厅的目标和所展示的企业核心文化目的是什么。通过创意与概念、整体思路与表达、展示设备与展区设计等过程形成合理逻辑推理，然后通过组合空间、材质应用、交通流线布局、多媒体运用，最终形成设计上闭环。在整体逻辑构架、逻辑表述及汇报 PPT 等节点上，应该是串联起一个清晰的路线。该汇报的内容在逻辑构架方面上有所欠缺，但是在建筑空间构架与多媒体艺术的结合上，方案既综合利用了空间，也集成了海量的数据，最终形成互动结果，这是一个亮点呈现。

郭海

（1）绿色环保的主题，在此基础上可以表述得更深入，例如更加地突出可持续性，降低能耗等。

（2）两个方案都缺少对项目背景的表述，同时品牌推广（比如 Vis& 品牌文化的展示）也有所忽略，但这是极其重要的一个环节。

（3）通过市场的调研分析需要挖掘整理出参展商的具体需求，例如产品行业境况、功能性规划、材料环保应用、支持资金以及展馆搭建受限条件（比如限高）等。

(1) The proposition is of vital importance. The student shall first adopt the solution mindset to deduct from the proposition.

(2) The subject contents can be thoroughly analyzed in the aspect of functions in the design.

(3) Under the premise of green building, the design result not only shall be expressed in technical level, but also shall establish more extensive concept. It is a technical issue as well as an issue of idea. What influence will this idea have on the design logic structure? What's important to the student in this question is to first establish a system.

To judge the conception of life pulse, the shape similar to a capsule is formulated. In the deduction from life pulse to capsule, the final result developed from this form does not matter the most. In the contrary, what's the most important is the process of deducting from life pulse to capsule. Otherwise, the deduction can lead to many different results. It can be seen from the defense statement whether the life pulse is indicated as yin and yang from the conception to the conclusion. The design conclusion involves many concepts, such as space capsule. What's more, as for formal representation, how is the conception of life pulse represented in the form? Then the relation between the form and exhibition contents is established to thus form clear logical relationship. For example, what is set in the position of yang, and yin? In summary, the explanation process of the students is impressive to interconnect the universe with everything, further combine the space through scientific and technological analysis. This method is excellent. As for the design, the student shall consider how the space is created after the form is completed, and how the activity scenario space created from the enclosure is divided from the structure and defined?

Representation design is highly complicated in the scope of interior design through requiring the clear confirmation of the purpose of exhibition hall and the purpose of the core corporate culture demonstrated. The process of creativity and concept, general mindset and expression, exhibition equipment and exhibition area design, etc. develops the reasonable logical reasoning, and finally forms the closed loop of design through combined space, material application, traffic flow layout and multi-media application. The nodes of general logical structure, logical expression and report PPT, etc. shall be connected to form a clear route. The contents of the report are a little bit flawed in logical structure. However, as for the combination of building spatial structure and multi-media arts, the scheme not only comprehensively uses the space, but also integrates plenty of data to finally achieve the interaction result. This highlight is represented.

(1) The theme of green and environmental protection can be more thoroughly expressed on the existing basis, such as more vigorously highlighting sustainability and reducing energy consumption, etc.

(2) The two schemes lack the expression of project background, and ignore the brand promotion (i.e. exhibition of Vis& brand culture), which is actually very important procedure.

(3) The specific demands of the exhibitors shall be explored and sorted out through market survey and analysis, such as product industry conditions, functional plan, material environmental protection application, supporting capital and exhibition hall construction restrictions (i.e. hight limitation), etc.

答辩评审

高 校：浙江工业大学
College: Zhejiang University of Technolog
学 生：沈令逸、张毅津
Students: Shen Lingyi, Zhang Yijin
指导教师：吕勤智、王一涵
Instructors: Lü Qinzhi, Wang Yihan
课题分数：85
Subject Scores: 85

基于一体化设计与装配化建造的阿里巴巴青年公寓室内设计研究

Interior Design Research of Alibaba Youth Apartment Based on Integrated Design and Assembly Construction

沈令逸

张毅津

学生感悟
Student's Thought

在住房开发中，从多维的角度看，新型城市里的人们有独特的需求，预算和生活方式。在考虑居住方式时，他们中的大部分人都面临着冲突，探索这些问题并提供解决方案。分别是社会动态模拟，建筑室内设计与装配化建造。

全球化的背景之下，城市中的年轻人，尤其是阿里巴巴青年员工这样的科技型人才，他们有着多样化发展的工作和生活模式，他们渴望更多的社交和对外交流，所以他们既需要更私密的个人独立空间，又需要更开放的公共交流空间，且这种趋势会随着信息时代的到来更加显著。

人体尺度与空间分析 ❙ Analysis of Human Scale and Space

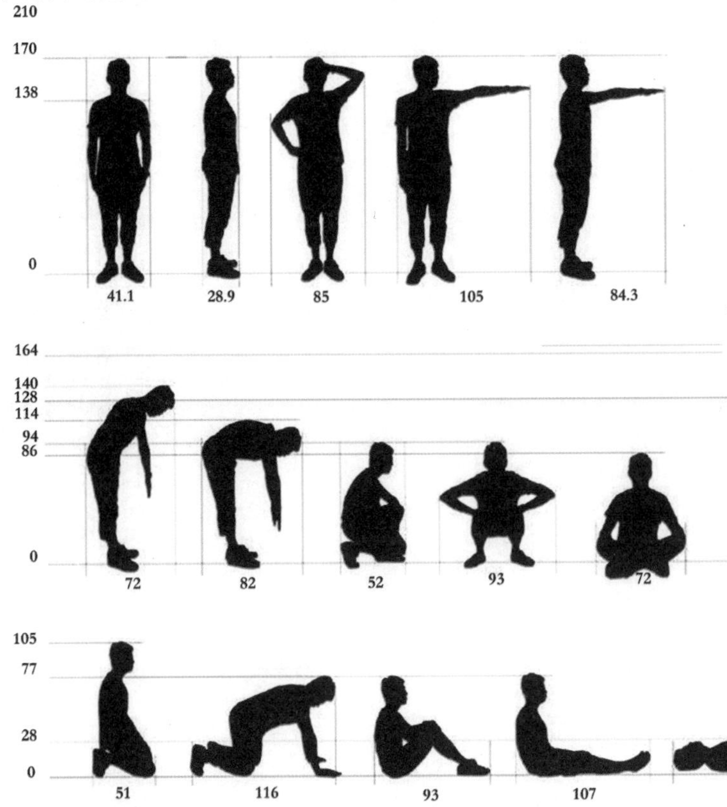

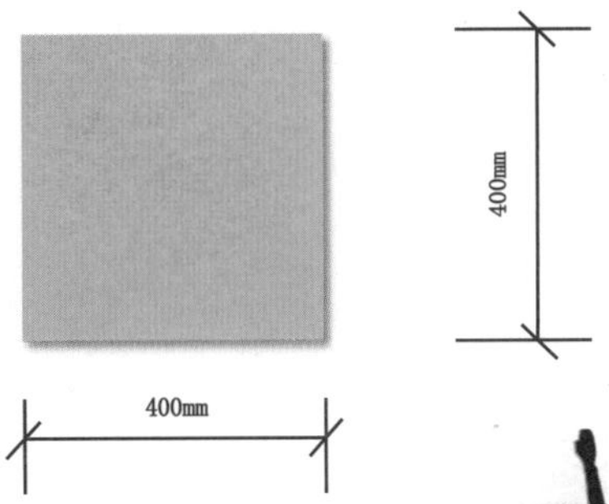

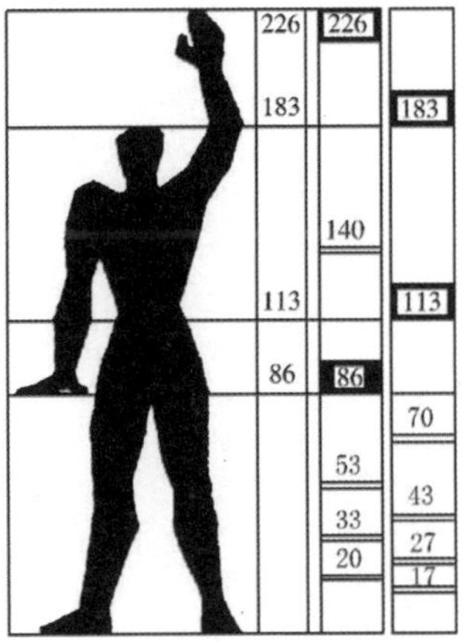

人体模度比例

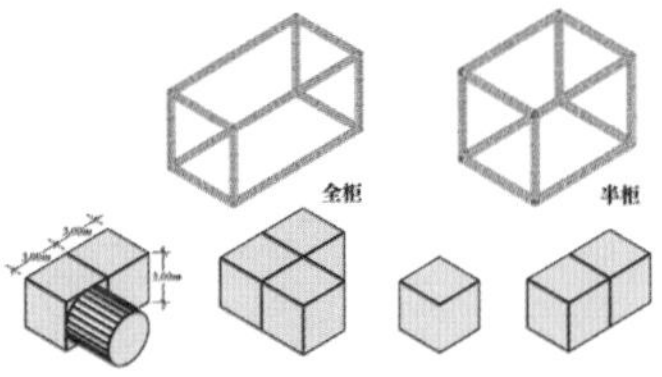

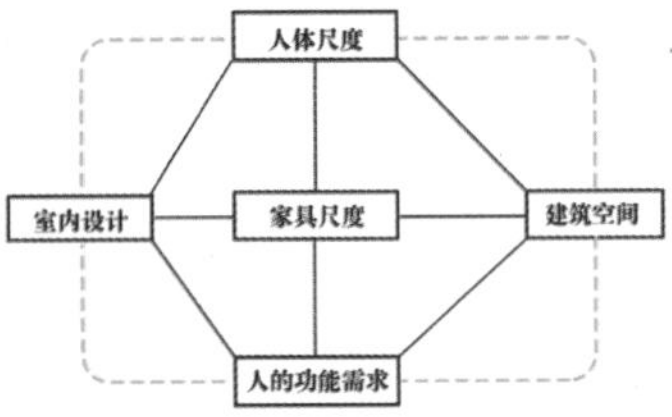

全柜　半柜

人体尺度

室内设计　家具尺度　建筑空间

人的功能需求

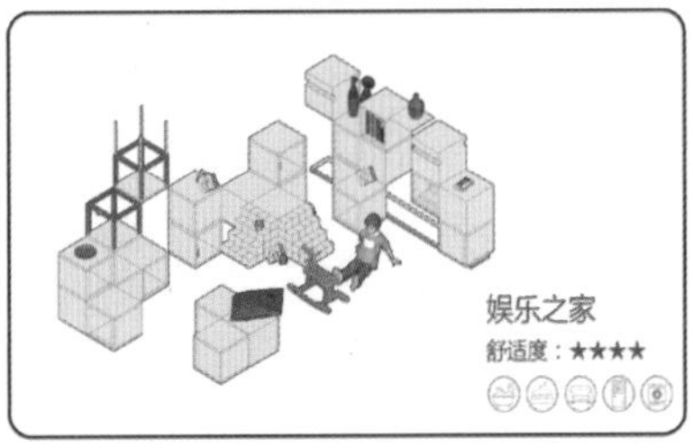

娱乐之家
舒适度：★★★★

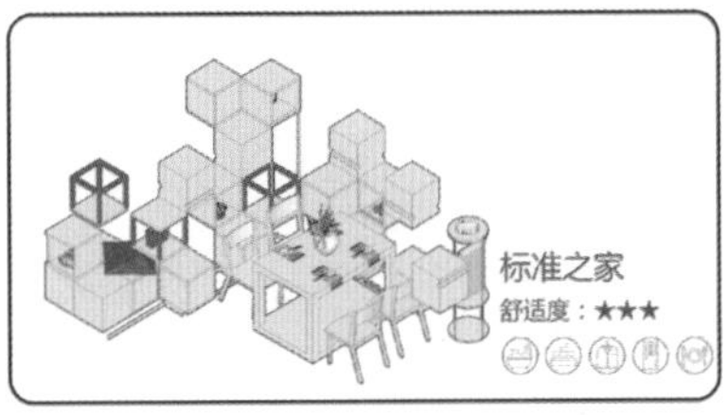

标准之家
舒适度：★★★

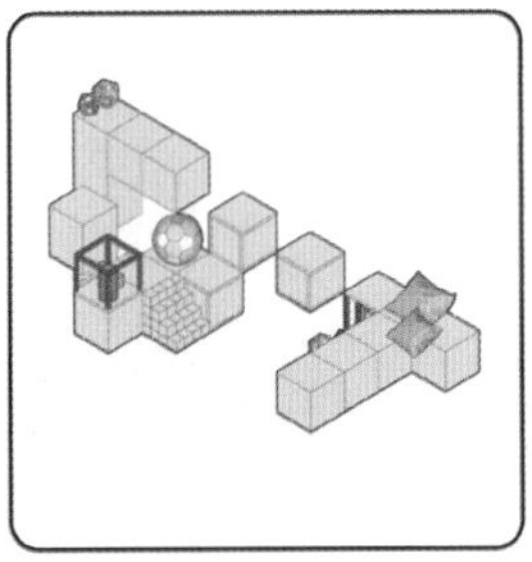

儿童区 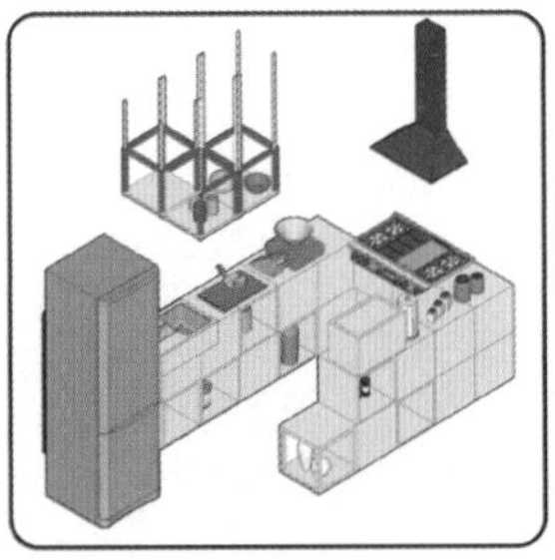　　厨房区域 　　书房区域

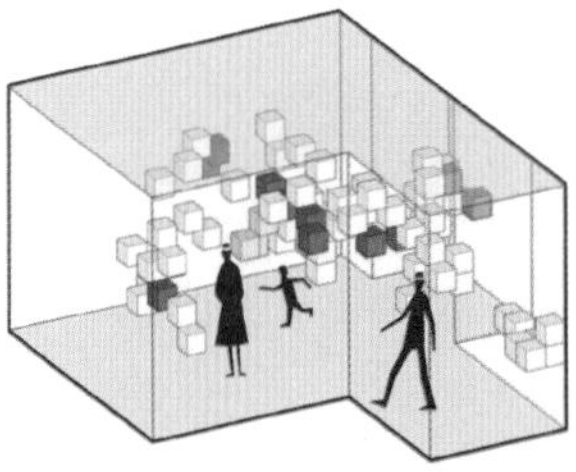

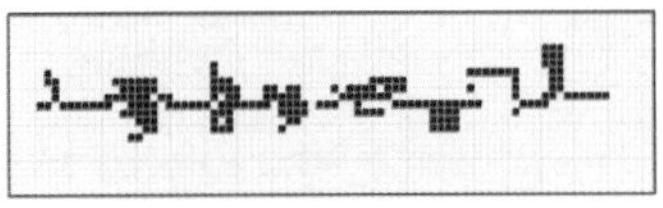

初始状态 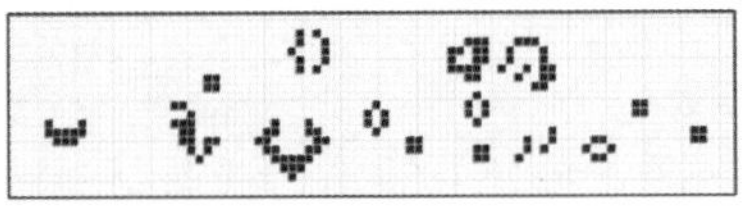　　稳定 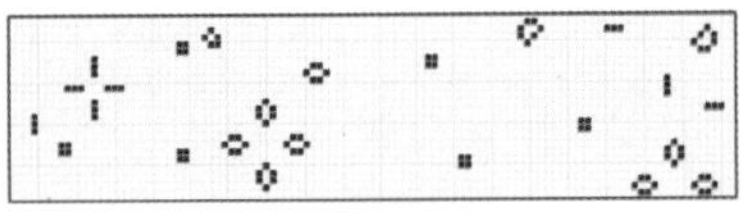　　熵变

153

绿色营造卷——建筑室内装配化设计

答辩评审

模块选材系统 ┃ Module Material Selection System

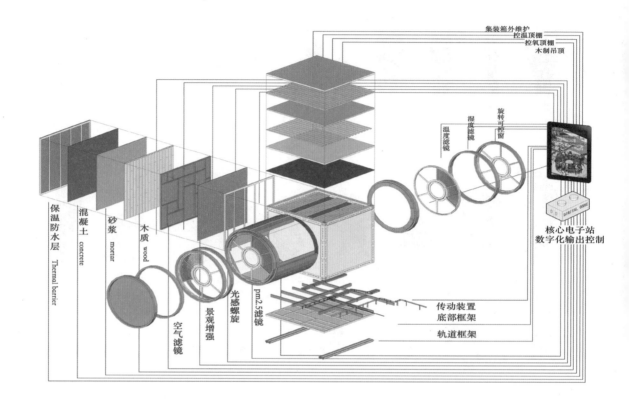

集装箱外维护
控温顶棚
控氧顶棚
木制吊顶

旋转可控窗
湿度滤镜
温度滤镜

核心电子站
数字化输出控制

保温防水层 Thermal barrier
混凝土 concrete
砂浆 mortar
木质 wood

空气滤镜
景观增强
光感螺旋
pm2.5滤镜

传动装置
底部框架
轨道框架

家具模块选择系统 ┃ Furniture Module Selection System

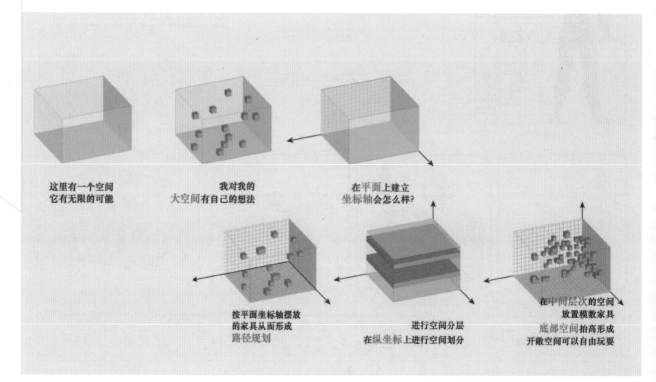

这里有一个空间
它有无限的可能

我对我的
大空间有自己的想法

在平面上建立
坐标轴会怎么样？

按平面坐标轴摆放
的家具从而形成
路径规划

进行空间分层
在纵坐标上进行空间划分

在中间层次的空间
放置模数家具
底部空间抬高形成
开敞空间可以自由玩耍

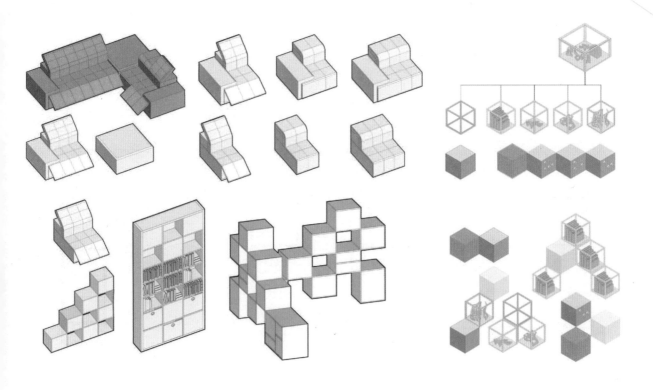

组团模块选择系统 ┃ Group Module Selection System

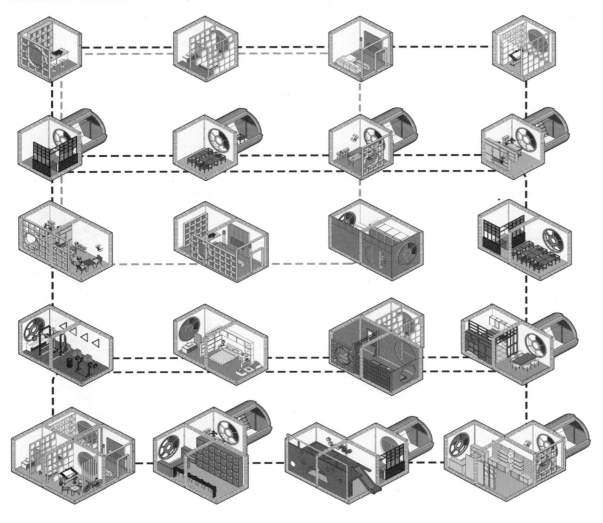

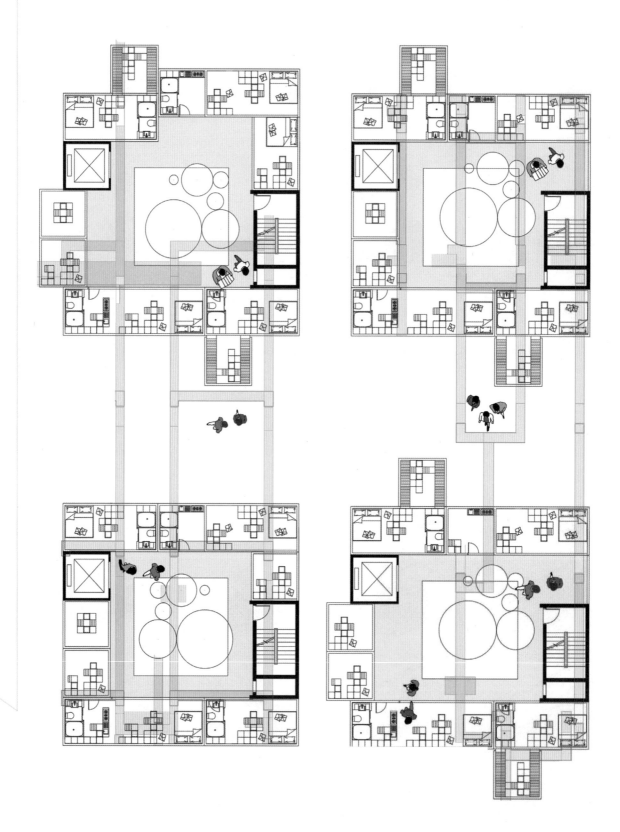

「室内设计 6+」2019（第七届）联合毕业设计
"Interior Design 6+"2019(Seventh Year) Joint Graduation Project Event

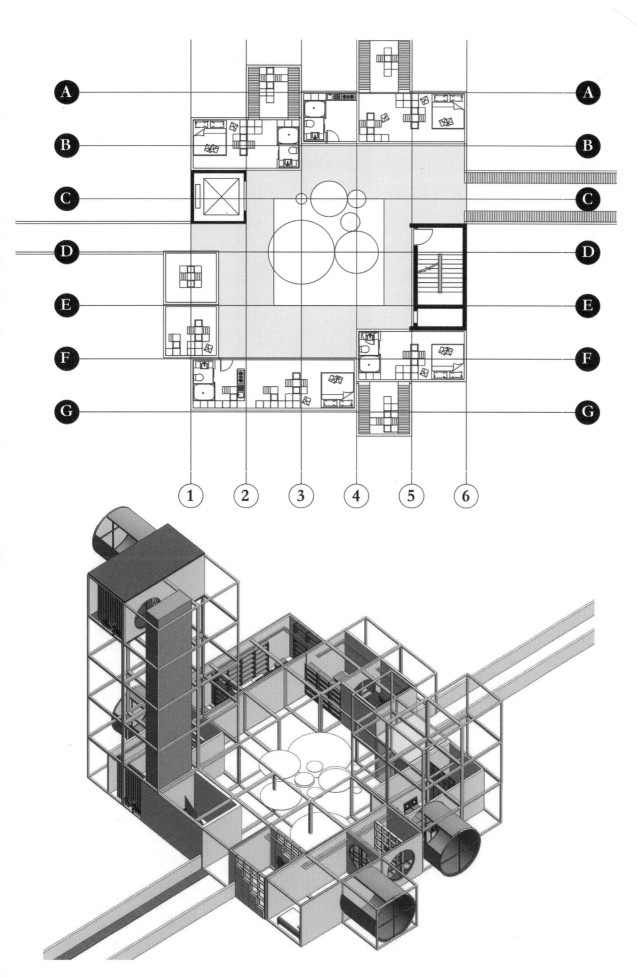

气象数据分析 Ⅰ Meteorological Data Analysis

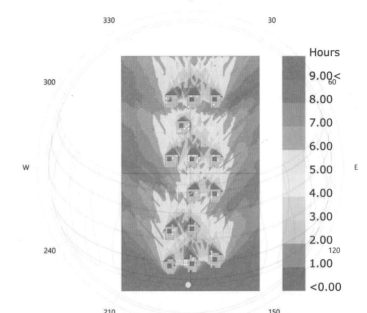

效果呈现 Ⅰ Effect Presentation

SunlightHours Analysis

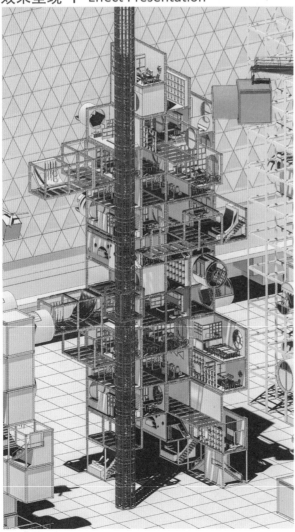

『室内设计 6+』2019（第七届）联合毕业设计
"Interior Design 6+"2019(Seventh Year) Joint Graduation Project Event

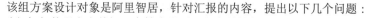

该组方案设计对象是阿里智居，针对汇报的内容，提出以下几个问题：

（1）如何体现人文关怀。在技术层面，方案做的较多，阐述得到位，但是所有的设计无论是住宅设计还是其他设计，都是要关注人文关怀，把人放在第一位的，以人文本。

（2）智能家居体现在何处。针对整个楼的物业管理状况，可能需要一个集中起来的空间作为共享空间。本人曾经参观过一个日本设计师伊东丰雄在台中设计的"天空墅"，20多层。它的顶楼设置了一个区域，所有的业主都可以使用。假如你的朋友来做客，家庭聚会你可以自己亲自下厨或者请厨师过来，也可以让物业安排。还有是不是需要打造并拥有一个喝下午茶的空间？

姚渊明

在该方案展示的过程中，阐述有理有据，大到整个课题宏观的阐述，小到设计细节的剖析，包括用一种舞台剧角色扮演的方式来说明方案的设计特色，都让答辩现场显得轻松有新意，整体表现出色，表述逻辑性非常强！课题虽然像是企业定制，但却十分难做，因为涵盖的东西太庞杂。无论是研究阿里的定向还是举一反三地做其他项目落地，智能的家居诸如用手机操作电器开关等等，都只是一种方式呈现，更多突出的应该是人性化设计。该组最后的方案在一体化设计中，最重要的人性化方面突出较少。阿里的程序员每天上班都够辛苦，如果下班仍在这样"人情味"不足的方盒子环境下起居生活，即使看似设计一体化了会不会感受上更郁闷？所以整个课题在"方盒子"的限定和方式下，在使用者的心理体验和人性化设计方面应该更多地去深入考虑，那么方案就会更加饱满和真正"一体化"。

马军

在院校构图的背景下，同学们和老师们往往会关注设计感的容量而忽略技术和建造的内容。这会促使并推动我们转移注意力并把重心放在同类以外的技术上。因此，每个历史时代出现了新技术或新思维的时候，我们会不留余力去追求它，那么技术将会发生偏移，我们也会忽略设计中最本质原始的东西。如果空间中最重要的是关注人本身，这个成果很大程度会变成产品设计，关注个体。当这种技术变成了具体的物品，物品的组合一起变成建筑物或生活中的空间场景，对个体的关注就会大大减弱。

谢冠一

The project design object of this group is the smart residence in Ali, and the following questions are raised for the contents of the report:

(1) How to show humanistic care. At the technical level, schemes have been done a lot and the elaboration needs to be well achieved. However, all designs, no matter residential design or other designs, shall pay attention to humanistic care, put people in the first place with the people-oriented concept.

(2) Where is the smart home embodied? According to the property management status of the whole building, a centralized space may be required as a shared space. I once visited a large Villa in the Sky designed by a Japanese designer, Toyo Ito, in taichung, which is over 20 floors in total. It has an area on the top floor that all the owners can use. If your friends come to visit, during family party, you can cook yourself or ask the chef to come over, also can let the property arrange. And is it necessary to create and own a space for the afternoon tea?

In the process of the presentation of the scheme, the explanation was reasonable and justified, ranging from the macroscopic elaboration of the whole subject to the analysis of the design details, including the use of a stage play role-playing way to illustrate the design features of the scheme, which made the scene of the defense seem easy and innovative, with excellent overall performance and strong logic expression! Although the topic is like enterprise customization, it is very difficult to carry out because it covers so many things. Whether it is to study Ali's orientation or to make other projects by analogy, the smart home such as using mobile phone to operate electrical switches is only a way to present. And the humanized design shall be more prominent. The final scheme of this group has less emphasis on the most important aspect of humanization in the integrated design. Ali's programmers are hard enough to go to work every day. If they still live in such a square box environment with insufficient" human touch" after work, will they feel more depressed even if the design seems to be integrated? Therefore, the whole project shall be given more in-depth consideration in terms of users' psychological experience and humanized design under the limitation and mode of " a square box", so that the scheme will be fuller and truly" integrated".

In the context of the institutional composition, students and teachers tend to focus on the capacity of the sense of design and ignore the contents of technology and construction, which will prompt and push us to shift our focus to technologies other than our own. Therefore, there's a new technology or a new idea in every historical era, and we will spare no effort to pursue it. Then the technology will shift, and we lose sight of the essentially original ones in the design. If the most important thing in the space is to focus on people themselves, the result will largely become product design, focusing on individuals. When the technology becomes concrete objects, the combination of objects becomes a space scene in a building or life and the focus on the individual is greatly reduced.

小户型租赁住宅装配化设计

Assembly Design of The Small Rental Housing

高　　校：同济大学
College: Tongji Unversity
学　　生：佐藤辉明、肖晓溪、王安琪、华立媛
Students: Sato Terumi, Xiao Xiaoxi, Wang Anqi, Hua Liyuan
指导教师：左琰、林怡
Instructors: Zuo Yan, Lin Yi
课题分数：90
Subject Scores: 90

学生感悟
Student's Thought

佐藤辉明

　　我们小组设计出来的东西和讨论的问题相对更细节细心一些。我深深体会到团队合作的不易和每个人都有属于自己的特长，当找到每个人特长的正确使用方法时，整个团队就会变得很强大。很有成就感很开心很满足！

肖晓溪

　　本次课题让我对室内装配化设计有了更深入的理解，它不仅仅是从家具单方面的设计，而是从建筑整体为出发点进行设计。感谢老师整学期的耐心指导，以及小组成员的团结协作，希望在未来的日子里，大家一起进步。

王安琪

　　通过一个学期的学习，我对室内设计有了更深入的了解，也体会到了团队协作的挑战与乐趣。感谢左老师林老师的谆谆教导，也感谢小组成员的通力合作，希望在未来的学习和生活中，持续吸收养分，不停止追寻进步的步伐。

华立媛

　　这一个学期我踏入了一个前所未有的新领域，从这个室内设计改造项目中，我通过各个环节了解了该如何进行实地调研和市场研究来做好一个有意义、有意思的室内设计。通过这个课题，我学习到了"小户型设计"和"装配化室内设计"的知识，了解到了室内设计领域最前沿、社会中最热点的话题。

162

外立面改造 ‖ Facade Reconstruction

立面设计 ‖ Facade Design

地面灯
LED 功率：13W
控制方式：常亮
防护等级：IP67
固定灯珠颜色：白色

晴天见
On a Sunny Day

公共空间设计 | Public Space Design

功能分析

原公共空间平面   梯间区域   现沿街商铺   现公共空间

晴天见 ON A SUNNY DAY

8	R801 - R804
7	R701 - R704
6	R601 - R604
5	R501 - R504
4	R401 - R408
3	R301 - R308
2	R201 - R208
1	? 服务中心 SERVICE CENTER
	休息处 RESTING AREA
	卫生间 TOILET

材质：铝氧化＋烤漆
标识字体：造字工房悦圆常规体
备注：黑色颜色
RGB：47、47、47

方案设计 1
Program Design

　　将快递柜，接待处放在门口，方便管理，也可以保证公寓的安全问题，可以随时登记进出入口。

　　平面尽头设置了厕所，楼上每户都只有一个厕所，在遇到厕所高峰时，住户可来这里借用。

　　活动室是用玻璃折叠门和外面相隔，住户可以在这里聚会、用餐。也可以将玻璃折叠门打开，这样就是一个完全敞开的公共空间。

公共空间平面图

连枝 A1 户型—情侣

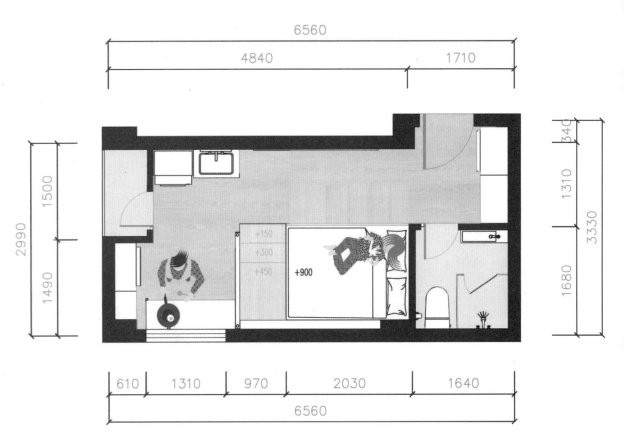

6560

4840　　1710

340
1310
3330
1680

2990

1500

1490

+150
+300
+450
+900

610　1310　970　2030　1640

6560

连枝 A2 户型—情侣

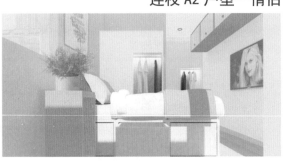

「室内设计 6+」2019（第七届）联合毕业设计

"Interior Design 6+"2019(Seventh Year) Joint Graduation Project Event

木兰户型 B

房型 B1

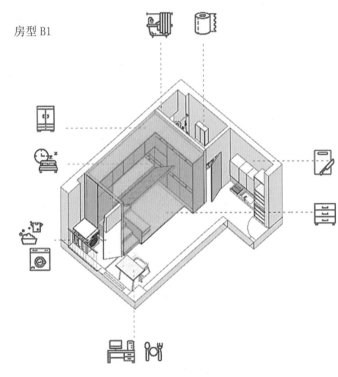

房型 B2

房型 B3

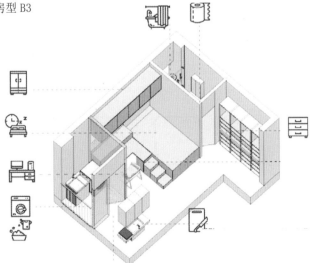

朝阳户型 D

房型 D1

设计策略：
打造拥有独立空间，公共活动空间，面积虽小，其乐融融的三口之家住宅。

户型：D-D1
面积：35 m²
所在楼层：5～7 层
总户数：3 户（9%）

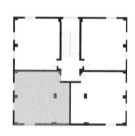

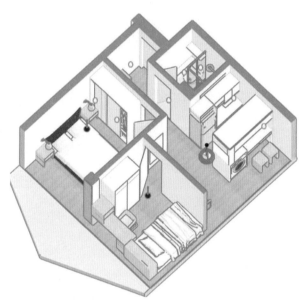

房型 D2

客群占比：
适应三口之家，约占总租客中 10%～12%
（总户型中占比：6/36 户，17%）

户型：D-D2
面积：35 m²
所在楼层：2～4 层
总户数：3 户（9%）

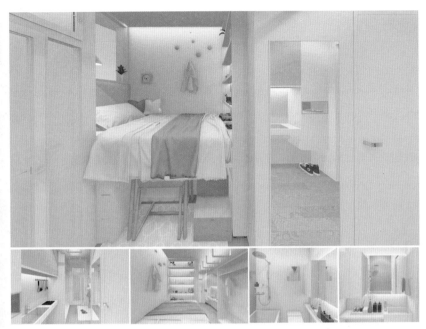

梦美户型 A

客群定位：

　　在深圳白石洲附近区域工作奋斗的单身人士。刚踏入社会，收入还未达到理想目标。住房预算不高，但也渴望独居生活。工作繁忙，偶尔把工作带回家中，作息时而不规律。想要回到家可以好好休息，睡个好觉。

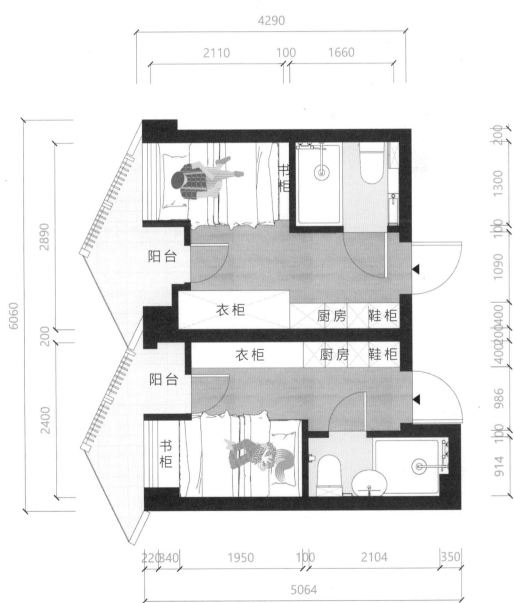

王传顺

该组分了 4 个不同的类型，切合深圳地方的文化和需求及刚需租赁的情况。整个设计由里及外考虑得都很详细，平面上让台的部分超出去 6m 宽，三角形部分超出去 1.7m。另外，建筑效果图的外立面是多彩的并且错落有致，把平庸的立面改造的新颖。多功能考虑设计非常好，利用整个外立面形体，方案设计用心细化了，其中具体的平面分析也蛮好的。这里有几个注意点：

(1) 平面功能方面做主要的。

(2) 大量运用家具组合，靠装配式来解决的前提是对的。因为面积不大，高效落地才能去实现目标。

(3) 学习考虑分析精炼的家具并把它们提炼出来是很重要的。房间如何显得比较大，关键还是看家具的组合。

(4) 如何做到花钱最少，效果最好。例如，像室内外建筑的外立面一样，利用色彩形成室内氛围，让不同的使用人群对于打造建筑室内外空间环境效果的色彩有着不同的感受。

(5) 功能细节方面需要琢磨。比如说浴缸，砌白石临时放进去是不是高了一点，泡浴缸的尺度上面有些地方需注意。

(6) 整个组合装配式的灵动性还是不错的，方案详细，但落地性要继续探索。在小房间里这么小的面积当中，应该注意把创意和生活中的实用性结合起来，得到社会和市场的认可，这是双收获。

刘恒

从整体性来考量事物的方法，细节方面很好。再从人以及内部的细节反推出来。题目的设计切入点很精准，居住区改造、居住人群的需求、使用习惯和尺度逐渐来研究出内容。这是比较好的解决程序问题和人的尺度问题的方案。另外，根据日照找到立面生成的一些方法体系，这个方法不错，达成了一个有组织逻辑、从自然中找到逻辑。该三角形语言在立面处理经常能见到，但有时候会出现和功能使用相冲突的矛盾性。在形式化的表达以后，要考虑有多少地方可以使用，等真正到了住宅的时候能卖多少钱。这个转换的过程中会有切分点，关键在于如何处理。使用的相关要求只会带来形式上的变化而已。和上一个方案相比，感觉该组方案实用性很强，具有落地性，可以执行实施。上午同题目设计方案的设计整体性和概念性比较好，建筑、装饰和家具一键化解决，是个好方案。个人认为，在企业化小空间的时候，越挤越好，手法少一点比多一点好。这意味着用简单的方式能解决更多的问题。临摹可变、可拆卸和可转换，那么将来的利用率会更高。而亮点是学生现在能考虑到运营的问题，这是十分难得的。从设计到运营，离不开计算造价。很多项目是从投资开始的，和甲方一起去做。这样能有新的不同的视角看待事情。除租赁外，运营成本、使用成本（包括水电成本、物业成本等）甚至遮阳处理是否能带来很好的提升。像其他能源的利用，可能又有一个新的亮点。

刘晓军

本方案是针对当下深圳年轻打工一族而设计的集成式小户型居住空间，重点关注低收入人群的居住品质，是极具现实意义的选题。整体方案设计完整、分析到位、逻辑清晰、图纸完成度高。建筑外立面、公共区及住宅内部都洋溢着时代的气息，针对不同居住人群设计的多个户型，同时可满足单身、夫妻、合租等不同人群的需求。居室内有效利用每平方米的空间，重点设计了具有可变组合功能的集成家具，其造型、尺度、功能都充分满足居者就寝、活动、收纳等日常生活需求。线条简洁的原木色家具和白色墙面以及个性化灯具的搭配不但把小空间的视觉感变得宽敞，同时让我们看到了家的舒适和温馨。

本案集成式卫生间的设计是值得肯定的，但应细化墙板以及管道的尺寸，特别是 A 户型中 11m² 和 6m² 超小面积中卫生间的设计。

The group divided the design into four types, which are in line with the local culture and needs of Shenzhen and the need for rental. The whole design was considered in detail from the inside and the outside. On the plane, the part of the table is beyond 6 m wide, and the triangle part is beyond 1.7 m. In addition, the facade of the architectural renderings is colorful and well-proportioned, transforming the mediocre facade into a novel one. Multifunctional design is very good, using of the whole facade shape, the plan design is carefully refined and the specific plane analysis is quite good. Here are some points:

(1) Plane function is the main function.

(2) It is wise to extensively use the furniture combination and solve them by assembly. Because the area is small, only through efficient landing, can the goal be achieved.

(3) It's very important to learn to think about analyzing refined the furniture and refining it. And the combination of the furniture will decide that whether the room can appear bigger.

(4) How to spend less and get the best results. For example, like the facades of indoor and outdoor buildings, colors are used to form the indoor atmosphere, so that different users have different feelings about colors that create the effect of indoor and outdoor buildings space environment.

(5) Functional details need to be pondered. For example, the bathtub, whether the masonry white stone temporarily put in is a little higher, there are some places on the scale of the bubble bathtub that we shall pay attention to.

(6) The whole combination assembly is of good flexibility and the scheme is detailed, but the implementation needs to be further explored. In such a small area of a small room, we shall pay attention to the combination of creativity and practicality in life, and get the recognition of the society and the market, which will bring a double harvest.

The way to consider things from the perspective of wholeness is very good in detail. Then we can deduce it from people and the details inside. The design entry point of the topic is very accurate. According to residential area transformation, the needs of residents, use habits and scale, we will gradually study the contents. This is a better solution to the problem of procedures and human scale. In addition, we find some method system of facade generation according to the sunlight, through this good method, an organized logic can be achieved, with the logic found from nature. This triangular language is often seen in facade processing, but sometimes it conflicts with functional usage. After formal expressions, we consider how much space we can use and how much we will sell when we really arrive at the house. There will be cutting points in the process of this transformation and the key is how to deal with it. The related requirements of use will only bring about a change in form. Compared with the previous scheme, I feel that schemes of this group have very strong practicability and implementation and can be carried out. The design integrity and concept of the design scheme of the same topic shown in the morning are relatively good, and it is a good scheme to have the one-key solution for architecture, decoration and furniture. As for me, when the enterprise is in small space, the more crowded the better, and less technique will bring better effect than that with more techniques applied, which means that more problems can be solved in a simple way. Copying is variable, removable, and convertible, and the utilization rate will be higher in the future. However, the bright spot is that students are now able to think about operations, which is very rare. From design to operation, it is inseparable from the calculation cost. Many projects start with investment and are done with Party A. It gives you a new and different perspective to treat things. In addition to leasing, whether operation cost, use cost (including water and electricity cost, property cost, etc.) or shading treatment can bring a very good improvement. Like the use of other energy sources, there may be another bright spot.

This scheme shows an integrated small-sized living space designed for the young workers in Shenzhen, focusing on the living quality of the low-income people. It is a very realistic topic. The overall scheme design is complete, with full analysis, clear logic and high drawing completion. The facade of the building, the public area and the interior of the house are full of the atmosphere of the era. Multiple apartment types are designed for different residential groups, which can meet the needs of single people, couples, joint renters and other groups. In the bedroom, the space of every square meter is effectively utilized, and the integrated furniture with variable combination function is designed. It's shape, scale and function fully satisfy the daily needs of bedtime, activities, accommodation and so on of residents. The collocation of plain wood furniture with the concise line, white metope and personalized lamps not only changes the visual sense of little space capacious, but also makes us feel that the home is comfortable and sweet. The design of the integrated type toilet in this scheme is worth affirming, but the dimension of wallboard and conduit shall be refined, especially the design of toilet of the super small area in 11 m² and 6 m² of A house type.

高　　校：哈尔滨工业大学

College: Harbin Institute of Technology

学　　生：段然、张睿、谢雨萱、洪汉森

Students:　Duan Ran, Zhang Rui, Xie Yuxuan, Hong Hansen

指导教师：刘杰

Instructors:Liu Jie

课题分数：90

Subject Scores: 90

阿尔山西口村民宿示范区环境设计

Environmental Design of Demonstration Area for Villagers'Residence in AIshankou

172

段然

张睿

谢雨萱

洪汉森

学生感悟

Student's Thought

　　这次设计是我第一次真正意义上的小组合作，也体会到了小组合作的一加一大于二，明白了大家共同付出才会让这次毕设变得更有意义，在和老师组员们一起调研的过程中也激发了我设计的热情。

　　此次六校联合毕设为我四年的大学生活画上句点。从前期调研到答辩的一路奔波让我深切感受到作为一名设计师的辛苦和责任。同时毕设使我成长，懂得与成员之间的交流合作，相互间的碰撞和拼搏也都将成为珍贵的回忆。

　　在毕业设计的学习过程中，我学会也体会了很多，对于装配化施工的陌生到摸索，使我认识到唯有不断学习探索才能跟上时代的潮流。学习是一条漫长而艰苦的路，不能靠一时激情，养成良好的学习习惯至关重要，学习贵在坚持。

　　通过一个学期的学习，我对室内设计有了更深入的了解，并从更加专业的角度全面认识了民宿建设的重要性、建造过程以及未来改造方向，等等。感谢老师的指导和督促，也感谢小组成员的通力合作，希望未来还能够一起学习，共同进步。

场地设计 | Site Design

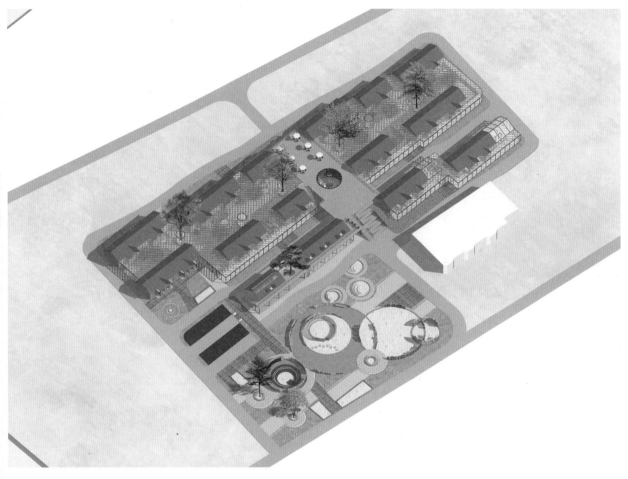

由于原有场地内游客的活动空间很少，我们将建筑南侧的空地扩大，作为游客和村民的室外活动场所。考虑到内蒙古寒地气候的特殊性，此场地的使用季节也是以夏季为主。同时，游客也拥有丰富的游览动线，场地内有多条人行道路，使各个场地之间相互贯通，游客可以在里面自由穿梭，产生互动和交流。也表达了聚合之意。

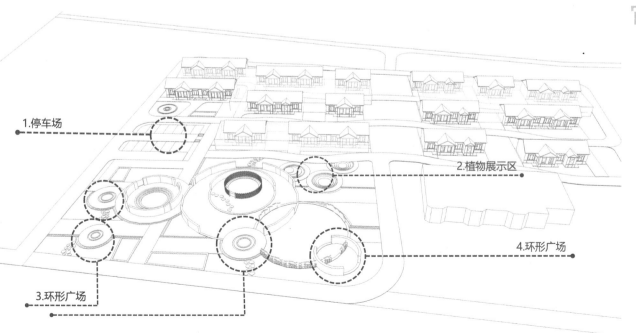

1.停车场

2.植物展示区

3.环形广场

4.环形广场

景观部分选用的植物均是就地取材，充分利用阿尔山地区丰厚的自然资源，这样不仅可以减少树种移植对生态的破坏，而且节省了运输成本。为营造丰富的景观层次，植被选择主要分为乔木、灌木、草坪三个等级。

使用需求：

种植体验

植物观赏

原有的交通线路混杂，人车不分离。因此我们对场地内的交通流线进行了重新规划。环绕场地一周的外部环线为消防通道，私家车则沿西侧道路进入停车场，再由北部道路驶出。场地内部仅供行人使用。我们通过人车分离保证了场地内行人的活动安全。场地内的环形小剧场在考虑到游客视线交错的同时还提供的多元化的使用功能，为游客提供种植体验、植物观赏、聚会闲聊、夜间观影等丰富的使用体验。

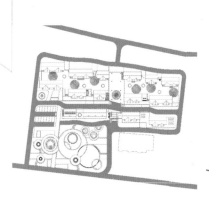

—— 消防流线

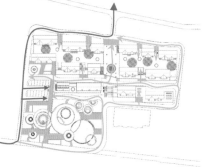

—— 人行流线
—— 车行流线

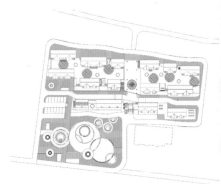

—— 绿植区域
—— 铺装区域

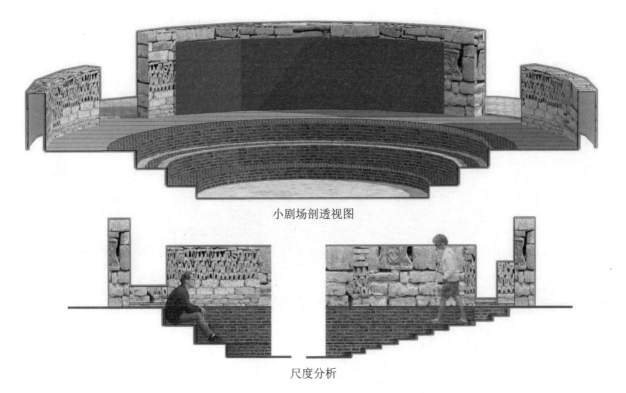

小剧场剖透视图

尺度分析

室内设计 ┃ Interior Design

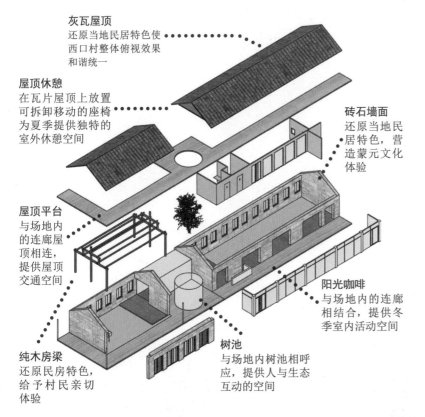

灰瓦屋顶
还原当地民居特色使
西口村整体俯视效果
和谐统一

屋顶休憩
在瓦片屋顶上放置
可拆卸移动的座椅
为夏季提供独特的
室外休憩空间

屋顶平台
与场地内
的连廊屋
顶相连,
提供屋顶
交通空间

纯木房梁
还原民房特色,
给予村民亲切
体验

树池
与场地内树池相呼
应,提供人与生态
互动的空间

砖石墙面
还原当地民
居特色,营
造蒙元文化
体验

阳光咖啡
与场地内的连廊
相结合,提供冬
季室内活动空间

活动方式分析

阳光咖啡区,人们可以在屋顶平台或室内休息区品尝饮品沐浴阳光,大面积的南侧开窗保证了充分的采光。

展览区可以通过折叠墙体转换为独立分割的3间活动教室,在旺季时人群可以自由穿梭,增加空间通透性。玻璃连廊尽头的公共休息区与大厅休息区相呼应,选用民族手工编织地毯与满族特色窗框,形成视线节点。

由于阿尔山地区旅游人口淡旺季数量变化明显,我们将公共服务区的部分墙体设置成可移动拆卸式的,在淡季外来游客较少时,原有的交通空间被折叠墙体隔开,变为3个活动教室。活动教室用于淡季给西口村村民学习生产生活技巧,将原有村社区内的教学课程转移到这里,给予村民更舒适便捷的学习环境,在充满民族氛围的室内装饰中熏陶艺术造诣,以便村民制作更加富有文化与美学内涵的手工艺制品。以此给当地村民提供新的收入来源,带动当地经济发展。

• 体块生成分析

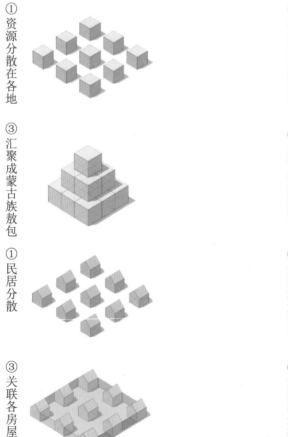

① 资源分散在各地

② 将元素分类整合

③ 汇聚成蒙古族敖包

④ 敖包组成聚落

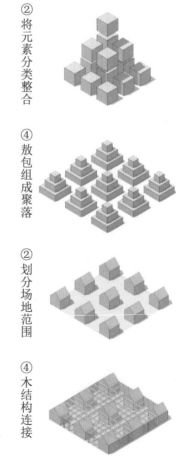

① 民居分散

② 划分场地范围

③ 关联各房屋

④ 木结构连接

　　我们在设计中提取蒙古族与满族民居特点，在场地内融入交汇与融合的设计主题。提取敖包的形态生成过程象征多种元素的整合，敖包是由分散在各处的石头堆砌形成的集会场所，敖包汇聚的地方便形成了部落与城镇。

传统满族民居的屋顶一般采用草屋顶和瓦屋顶两种形式。其中瓦屋顶一般用小青瓦仰面铺砌，瓦面纵横整齐。他与北京地区的合拢瓦不同，瓦片全部仰砌，屋顶成为两个规整的坡面，这主要是因为东北地区气候寒冷且冬季降雪量大，拢瓦旁的灰泥容易遭到积水侵蚀导致脱落。所以满族民居大多选用仰面瓦，主要是为了便于排水防止瓦片脱落。

项目中模仿当地传统民居的建筑手艺，将木梁与柱进行装饰化改造，在传承当地文化的同时还可以提升建筑稳定性。由于空间充沛，在三拼套房中，我们尝试设置满族传统民居中的口袋房、万字炕。

面对淡旺季变化问题提出双拼单元方案，旺季时作为两个单元房，淡季时摘除隔墙合并作为一个套房。

借助装配式隔墙系统安装的便捷性将不同时段下的空间充分利用。除去对满族传统民居"口袋房""万字炕"这种线条形式的探索外，在这次的室内设计中将传统火炕加以改造，融入日式榻榻米元素，使其在具有地方特色的同时还兼具舒适性。在民宿项目中将电视机移除，用小型投影仪来替代，在节省造价的同时还解放了空间。

淡旺季变化双拼户型保留了原有建筑的中间隔墙，将其一半改造为可折叠收纳的移动墙体。淡季使用人群较少时，墙体折叠收起，两个房间变为一个大套房，可供四人团队居住，餐厅合为一个，中间的公共区域变为开放起居室。旺季时墙体打开，分割为两个房间，每个房间均配备有厨房与餐厅，满足游客多种需求。床铺均选用火炕内嵌床垫的组合形式，让来自世界各地的游客都可以快速适应并获得良好的睡眠质量。

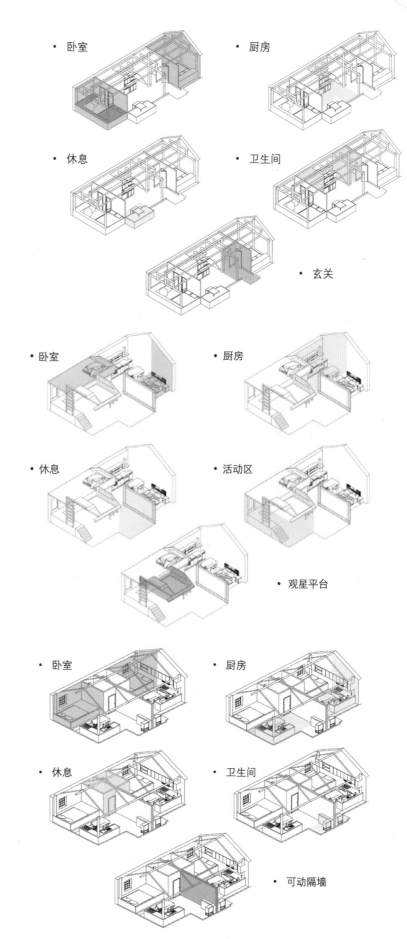

· 卧室　　　· 厨房

· 休息　　　· 卫生间

· 玄关

· 卧室　　　· 厨房

· 休息　　　· 活动区

· 观星平台

· 卧室　　　· 厨房

· 休息　　　· 卫生间

· 可动隔墙

177

徐敏

　　演讲很精彩，热点也很多。从设计制作的过程中可以感受到辅导老师花费了大量时间和精力。在设计中出现的满文，需要考虑满文翻译得是否准确，英文翻译是否正确，而非纯粹是因为觉得它好看。另外，本人发现城市里的美术馆，有时候建筑外墙上的英文字比中文文字还大。这个真的有必要吗？可能有同学会说，现在处于国际化的环境中，国外留学生有很多，所以要用英文来表示更为醒目。记得本人有一次参观清华大学美术博物馆，拍了照片回来，里面没有出现英文字比中文文字还大的情况，甚至广告牌上都无英文字。由此可见清华大学的设计观念。借助这个话题，针对满文的问题。本人想说方案的设计观念需要改变，不能完全只追求以形式为美的最终目的，然后把参数化和民宿结合在一起。

杨琳

　　本方案设计具有地域特色、民族元素和地标性、话题性。基地地处阿尔山全域旅游区域南部交通线上，是未来游客旅友理想的休憩地。同时，西口村也是文化和旅游部精准扶贫点，将持续打造成乡村旅游目的地。本方案设计定位符合乡村民宿的性质，为游客提供了体验乡村文化的环境设计，特别是蒙元文化和现在设计的结合，以现代设计诠释传统文化，并把功能和文化活动相结合，以环境为载体，呈现丰富多彩的艺术设计。圆形的下沉表演场隐含了蒙古包的建构元素；室内火炕火墙的改良更符合年轻人的使用；景观种植配置了当地丰富的植物物种，可以科普植物知识，温室的现代种植也为冬季游客提供了体验内容。优点众多、缺点是设计语言过多，不够概括兼容，需考虑乡村环境和建造成本控制，还要关注立体观景平台和整体性灰空间的建构防灾问题。

蒋浩

　　两位同学的表达充满着激情，其实设计也是需要激情的，不然太枯燥、无动力，整个设计的过程会感觉很累，我们的毕业设计更要有激情的去完成。在叙述中，关于设计的过程上要增加一些东西。例如在室外景观设计上，植物配置需要说明乔木、灌木及草丛的层次分布；在景观造型上，注意一些装饰的使用，尤其是景观照明方面的内容比较欠缺；在建筑和室内设计上，需要知道室内布局的合理性和功能划分。毕竟民宿不同于一般的星级酒店，它一定要有地域文化符号。把新文化融入到设计理念中，并将其结合起来。同时，两位同学所设计的景观建筑室内外一体化的结合很好，在设计上确保整个空间环境的统一和谐、人与自然的共融共生。

The speech is wonderful and there are many hot topics. It can be seen from the design process that the tutor has spent a lot of time and energy. I would like to make a suggestion here that Manchu appears in your design. Regardless of that whether the Manchu translation is accurate or not, whether the English is correct or not, the use of these words or English is just because they look good. In addition, I have found that in art galleries of cities, sometimes the English words on the walls of buildings are larger than the Chinese words. Is it really necessary? Some students may say, now the country is in the international environment and there are a lot of foreign students, so using English is to express more eye-catching. I remember that I once visited the museum of fine arts of Tsinghua University and took photos.There were no English words larger than Chinese words in the photos, even no English words on the billboards. This shows the design concept of Tsinghua University. With the help of this topic, the problem of Manchu, I would like to say that the design concept of students needs to be changed, and they can not only pursue the ultimate goal of beauty in form, and then combine parameterization with homestay.

The scheme is designed to show regional characteristics, ethnic elements, landmarks and topics. The base is located on the southern transportation line of Alshan regional tourism region, which is an ideal rest place for future tourists. At the same time, Xikou village is also the poverty alleviation point of the ministry of culture and tourism, and will continue to be a rural tourism destination. The design orientation of this scheme is in line with the nature of rural homestay. It provides tourists with the environmental design of experiencing rural culture, especially the combination of Mongolian culture and current design, interpreting traditional culture with modern design, combining functions with cultural activities and presenting colorful artistic design with the environment as the carrier. The circular sunken performance field implies the construction elements of the yurt; the improvement of indoor heated brick bed and fire wall is more suitable for young people to use; landscape planting is equipped with abundant local plant species, which can spread the knowledge of plants. The modern planting of greenhouses also provides the experience for winter visitors. There are many advantages. And disadvantages are as follows. There are too many design languages and not enough generalization and compatibility, which need to consider the rural environment and construction cost control, and pay attention to the disaster prevention of the construction of three-dimensional viewing platform and integrated grey space.

The expressions of the two students are full of passion. In fact, the design also needs passion. Otherwise, it will be too boring and unmotivated. And during the whole design process, we will feel very tired. Our graduation design must be completed with passion. In the narrative, something is added to the design process. For example, in outdoor landscape design, plant configuration needs to explain the hierarchical distribution of trees, shrubs and grasses; in the landscape modeling, we shall pay attention to the use of some decoration, especially lack of the landscape lighting content. In architecture and interior design, it is necessary to know the rationality of interior layout and functional division. After all, the homestay is different from the general star-rated hotel, it must have regional cultural symbols. The new culture needs to be integrated into the design concept and combined. Meanwhile, the landscape architecture designed by the two students integrates indoor and outdoor design very well, which ensures the unity and harmony of the whole space environment and the symbiosis of human and nature.

179

高　　校：北京建筑大学
College: Beijing Unversity of Civil Engineering Architecture
学　　生：乌云塔拉、马宇萌、魏懋榕
Students: Wu Yuntala, Ma Yumeng, Wei Maorong
指导教师：杨琳
Instructors: Yang Lin
课题分数：85
Subject Scores: 85

内蒙古阿尔山市明水河镇西口村幸福苑民宿改造

Reconstruction Project of Xingfuyuan Homestay in Xikou Village,Aershan City,Inner Mongolia

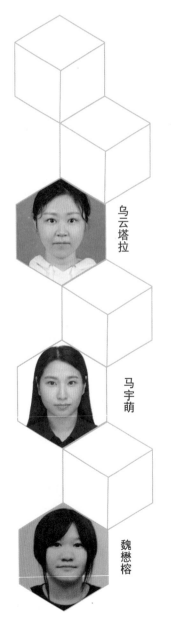

乌云塔拉

马宇萌

魏懋榕

学生感悟
Student's Thought

导师的教导耐心细致，"将建筑上升至文化"是我们这组设计的主旋律。在这次联合毕设中，使我明白与同龄人之间的差距仍然不小，这是让我以后无论从事什么行业，都避免懈怠的重要动力之一。

建筑室内设计的装配化趋势在近些年越来越普遍，在这次设计的过程中，我详细地了解了装配化和模数化在设计中的应用，并发现很多设计上的问题。这是一次非常有意思的尝试，感谢指导教师以及同学对我们的帮助。

第一次去实地调研并和小组共同探讨方案，以及多次与导师的协商让我渐渐领会到了这个领域真正的魅力。特别感谢老师教授给予的建议和帮助，也特别感谢小组同学给的鼓励和支持。相信在以后的道路上一定还会并肩同行更进一步。

民宿的沿革 | The Evolution of Homestay

学习苏联为中国建筑事业确立全面的管理体制奠定了基础，但也滋养了僵化的教条主义，滋养了以政治尺度度量创作问题的简单化做法，对建筑创作造成的影响是，思想的封闭禁锢了建筑创作的原动力。

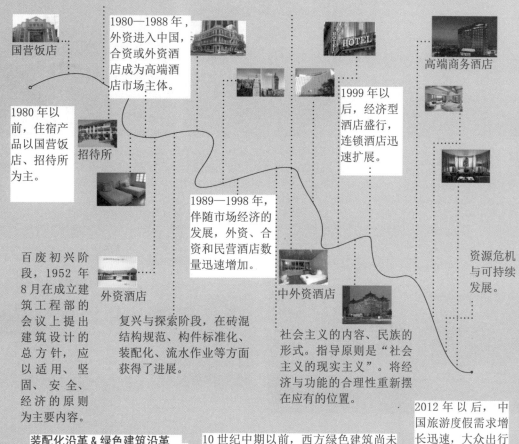

国营饭店

1980—1988 年，外资进入中国，合资或外资酒店成为高端酒店市场主体。

HOTEL

高端商务酒店

1999 年以后，经济型酒店盛行，连锁酒店迅速扩展。

1980 年以前，住宿产品以国营饭店、招待所为主。

招待所

1989—1998 年，伴随市场经济的发展，外资、合资和民营酒店数量迅速增加。

资源危机与可持续发展。

百废初兴阶段，1952 年8 月在成立建筑工程部的会议上提出建筑设计的总方针，应以适用、坚固、安全、经济的原则为主要内容。

外资酒店

复兴与探索阶段，在砖混结构规范、构件标准化、装配化、流水作业等方面获得了进展。

中外资酒店

社会主义的内容、民族的形式。指导原则是"社会主义的现实主义"。将经济与功能的合理性重新摆在应有的位置。

2012 年以后，中国旅游度假需求增长迅速，大众出行主体由商务出行转向个人旅游，居民对于客栈民宿等个性化主体酒店需求增加。

装配化沿革 & 绿色建筑沿革

发展初期：1950—1976 年，全面学习前苏联。

10 世纪中期以前，西方绿色建筑尚未有清晰理念，一些设计师开始意识到和谐处理建筑与环境的关系。

对节约钢筋、木材、水泥起到了积极的推进作用。

科学研究跟不上项目建设的速度，装配式建筑质量低劣。

20 世纪中期至末期，伴随环境问题的日益恶化，有关政府部门提出了生态建筑新理念。

高校先后成立混凝土预制专业

现浇混凝土的机械化。

发展起伏期：装配式建筑的停滞、发展、再停滞的起伏波动。

"慢生活"是西口村旅游发展的目标。"慢种慢养慢生活"意指"生产＋生活"，也即"乡村＋旅游"。乡村背后是没有完全实现的价值，比如知识价值等，这些价值暂时还没有变成"经济"价值，它需要某种转换机制。通过某种方式展示乡村生活的魅力、特色、品位，这些都会催生全新的、有活力的乡村经济。而乡村经济的活跃又会进一步带动人才、技术、资金、资源、信息等方面的流动与发展，直到达到城乡之间的互动平衡，实现乡村的逆袭。旅游就是帮助乡村实现价值的一种转换机制。

"四化、三改、两加强"，北方地区形成通用全装配化住宅体系。

现浇体系进入中国，预拌混凝土应运而生，建筑向高层发展。

居住建筑方面，城镇建设促进了预制装配式技术的应用。

答辩评审

区位分析 ┃ Location Analysis

  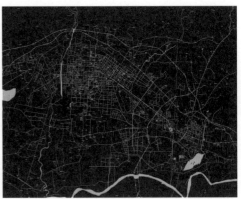

自然条件：西口村位于阿尔山最南端，气候寒凉，无霜期不足 100 天，难以发展大规模高效现代农业。但西口村处在农林牧三区交界处，地貌形态丰富，有优质的空气和水源，有充足的阳光和日照，具备发展旅游业良好的自然条件。

交通条件：西口村位于阿尔山市和乌兰浩特的中间。从西口村（明水河镇）开始，每隔 30～50km 的距离有一个镇或景点。两边向北延伸，和北部旅游环线相接，形成了南北旅游环线的贯通。

产业支柱：全市 6.8 万人，其中有 1.3 万人直接从事旅游，4.3 万人间接从事与旅游相关的行业。

发展旅游：目前，西口村是阿尔山最大的自然行政村，是全市贫困程度最严重、贫困面最大的地方。脱贫攻坚是当前最重要的任务。中央提出要打赢脱贫攻坚战，其目的不仅是脱贫，更要致富，是要全面振兴乡村，推动农业全面升级、农村全面进步、农民全面发展，让全体农民都走上可持续发展、共同富裕的道路。

民宿效果图

民宿效果图

西口村发展乡村旅游业的设想

制定一个中长期乡村旅游发展规划，为阿尔山岭南地区今后的长远发展明确定位。

建设西口村文化和旅游民宿示范项目。

打造京蒙协作"慢种慢养"生态农业产业园区，产业园区占地约 300 亩。

打造阿尔山最大的植物园。

建设明水河镇生态农业博物馆。

旧军用飞机跑道的保护和利用。

修建观光步行木栈和观景台。

修建观光自行车车道。

打造蒙元文化部落营地或博物馆。

建设停车场、农副产品销售大厅、旅游标志标牌系统、旅游厕所、公共区域路灯和监控系统等旅游基础设施。

规划分析

业态分析

【 室内设计 6+ 】2019（第七届）联合毕业设计
"Interior Design 6+"2019(Seventh Year) Joint Graduation Project Event

民宿功能定位 | Homestay Function Positioning

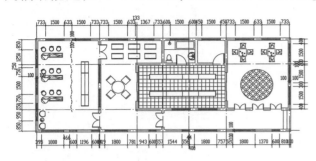

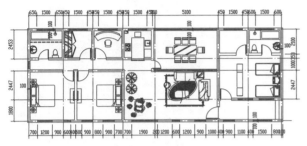

民宿居住空间平面图

项目名称："阿尔山房"内蒙古民宿建筑环境设计。

项目内容：原建筑园区占地 9000 m²，每套房间 7.2×5.1=36.72 m²，室内层高 3.0m；双拼 4 栋，3 拼 8 栋，共 16 栋 48 间，其中 4 拼 4 栋，可住 96～144 人。

建筑类型：民宿接待度假村。

建筑功能定位要求：

（1）住宿空间：满足一个套间和一个标间的体量，附有卫生间、更衣室、厨房、小客厅的基本配套设施空间。

（2）公共空间：布置展览区、工作区、休息区娱乐区，在此基础上加置手工艺品体验区。

建筑外观改造：根据现有外墙及屋顶出现的问题进行修整，进行绿色营造。

平面功能改造：原室内陈设欠妥，存在功能单一、空间多余和耗材浪费等问题，用装配式手法改正。

景观设计：满足绿色装配的应用标准和要求。

室内家具装配

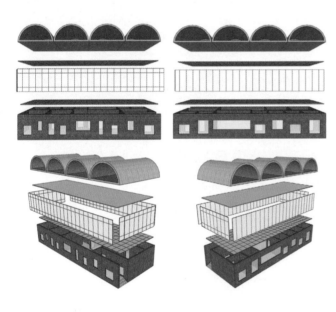

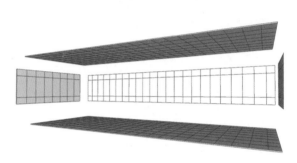

900mm×900mm 地砖与天花
900mm×3000mm（570mm+1560mm+570mm）墙体

建筑外观对应绿色装配选材 | Building Appearance Corresponds to Green Assembly Materials

绿化植物草皮

钢筋混凝土

清水混凝土砖

塑钢中空玻璃

作为坡屋顶的中和屋顶弧度参照蒙古人以天地为圆思考并根据俄式洋葱头屋顶简化延伸

纱网作为夹层给人以错高的假象是对帐篷的借鉴与应用

木制外墙保温，减法补充光线

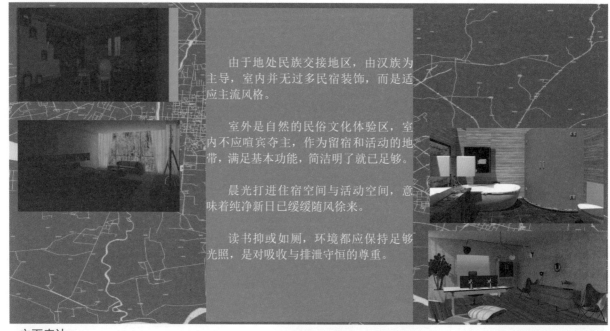

由于地处民族交接地区，由汉族为主导，室内并无过多民宿装饰，而是适应主流风格。

室外是自然的民俗文化体验区，室内不应喧宾夺主，作为留宿和活动的地带，满足基本功能，简洁明了就已足够。

晨光打进住宿空间与活动空间，意味着纯净新日已缓缓随风徐来。

读书抑或如厕，环境都应保持足够光照，是对吸收与排泄守恒的尊重。

立面表达

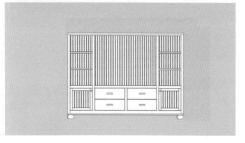

剖面示意

改造前后对比

民居改造自由性较大，拆除重建。原建筑运用普通毛坯上漆，风沙雨季寒潮容易脱落；室内布局单一但对称，家具与墙体老化破旧；由于受当地传统思想影响，只在南侧开窗，北侧设有厨房，由于密不透风加速墙体衰败。

在改造建筑中，将一层土房改建为二层小洋房，并以当地的木材与砖瓦为材料。考虑到要满足日常接待客人和用餐等需求，加设客房与餐厅。两层皆有绿化加布。

平面图均为示意图，一些陈设设有标注尺寸，由于不是采用同一软件制作，因此与效果图有出入。

公共空间轴侧展示

公共空间包括展示区域与休闲区域，其中休闲区域预留空间布置全息投影展示。展示内容分别为汉族传统建筑框架投影与蒙古包骨架投影，体现阿尔山地区多民族文化融合的特点。

居住空间轴侧展示

居住空间设置大床房与标准间，光纤穿孔天花板塑造星空效果，展现阿尔山美丽的自然风光，墙板由阿尔山远山影抽象而来，塑造良好的居住空间环境。

185

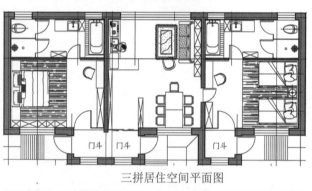

三拼居住空间平面图

三拼居住空间顶棚平面图

186

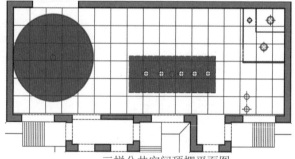

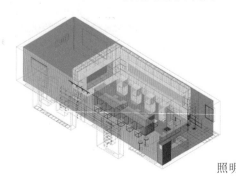

全息投影
展示预留
空间

三拼公共空间平面图

三拼公共空间顶棚平面图

照明类型分析图

休息区域
特殊照明

展示照明

重点照明

模块分析图

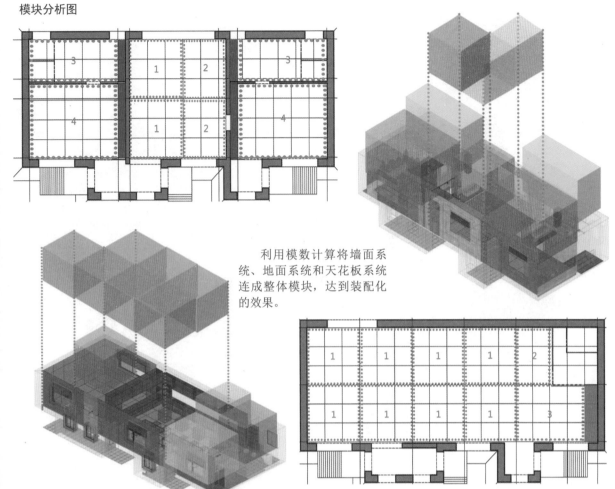

利用模数计算将墙面系统、地面系统和天花板系统连成整体模块，达到装配化的效果。

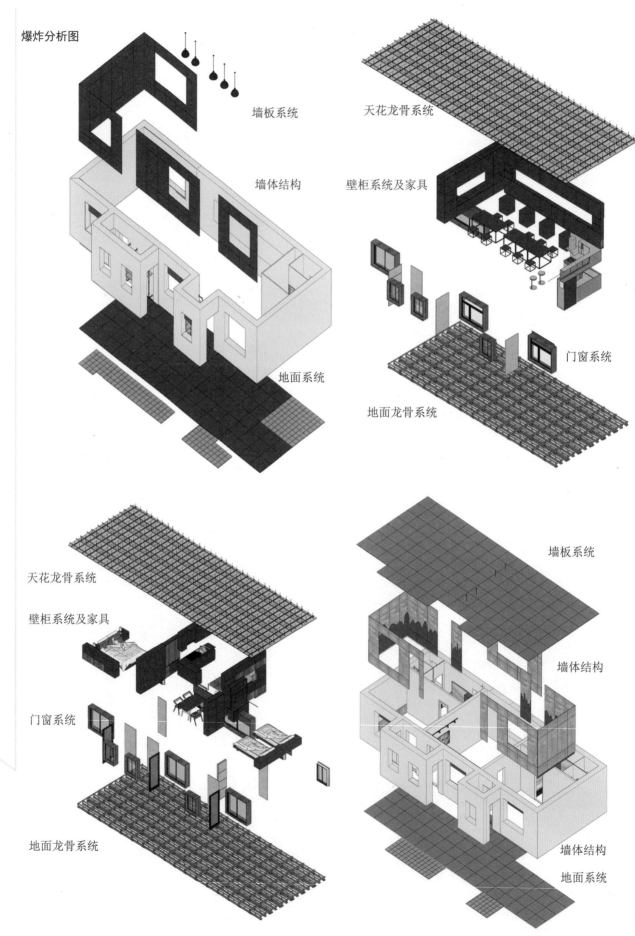

爆炸分析图

墙板系统

墙体结构

地面系统

天花龙骨系统

壁柜系统及家具

门窗系统

地面龙骨系统

天花龙骨系统

壁柜系统及家具

门窗系统

地面龙骨系统

墙板系统

墙体结构

墙体结构

地面系统

「室内设计 6+」2019（第七届）联合毕业设计
"Interior Design 6+"2019(Seventh Year) Joint Graduation Project Event

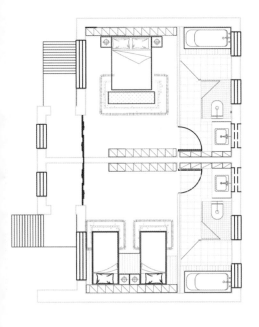

设计思路

在此次课题中，我们的设计重点在于对阿尔山房扶贫项目民宿的改造和相应的宣传。在政府的扶贫政策下，最初设想了很多开发方案，例如开发农业、畜牧业等，但是由于受地理位置和气候原因的限制，很多想法都无法实现。有人提出了一个发展旅游业的想法，在政府和专家的支持帮助下，修建立了近50间样板房。样板房分为三种户型（二拼80m²，三拼120m²，四拼160m²），此次分组中我选择了二拼80m²的样板间进行民宿改造和公共空间改造的设计。

阿尔山房当地的气候特点为冬天极其寒冷，在使用装配化的要求下，尽可能满足相应的保温要求和美观要求。此外为了吸引游客参观，我还添置了相应的与非物质文化遗产有关的设计和展示陈列空间，给游客带来更多的舒心和惊喜。

人体工程学分析 ｜ Ergonomic Analysis

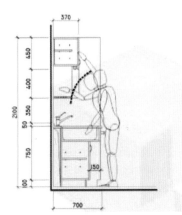

台面进深为700mm时，中柜底地面标高可降低，以增加储藏量

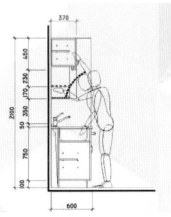

台面进深为600mm时，中柜底地面标高需升高，以防视线遮挡和碰头

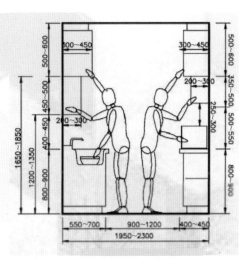

储藏空间内有富余空间是可增加柜台宽窄度

需要加高站立存取物品的高度范围可做成左右连通格，最大限度利用空间。
视平线位置站立存取最舒适的高度可以采用衣架等较方便的储存形式。
存取物品需要弯腰，根据使用习惯可做成格或抽屉。
存取物品需要下蹲或坐在床上的高度，可以做成抽屉，方便观察利于清洗。

① 镜面离地面的高度

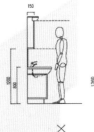

镜面离地面高度在1000mm左右为适中，若超过1200mm会给人造成视觉压抑感

② 洗手池上方柜门宽度

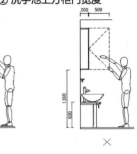

洗手池上方柜门不应过宽，在350mm以下更为合适，若过宽可造成开启不便

③ 水龙头与洗手池距离

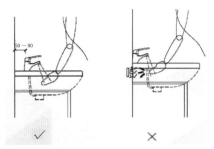

水龙头出口距离水池不宜过近，否则容易溅起水花且不易清洗

王传顺

该组方案的民宿设计别具一格，考虑十分详细，但需要注意的是，如何探索绿色和装配色的结合，突出彼此所具有的特色，并协调好它们之间的关系是至关重要的。就整个体系而言，尤其是空间组合这个版块，应该考虑将其运用到整个空间构造的大背景中。例如遮阳、墙体、吊顶以及家居的局部尺寸应是理性严谨的，要让它在空间结构的大环境里体现出来。虽然说设计师考虑的东西可能不是那么面面俱到，但可在后期补充详细的说明来完善方案的内容。个人认为，循序渐进、逻辑清晰的内容框架会更加好一点。因为最佳的逻辑是从大过渡到小，从整体到局部，由浅入深，而不是一开始就突出显示细节中的具体尺寸，这会给人一种中间部分体系的缺失感。

曹阳

该组方案主题性和中期相比越来越扣题。对于装配化的概念有了初步的认知与掌握。但地面为什么要架空，方案中没有明确表达原因。比如加厚墙板是为了加强墙体的保温性，架空地面是为了保障采暖性，加吊顶是为了解决空调和新风系统的布局。所以装配作为一种技术手段，在设计里还是显得有些生硬。

（1）从美感和舒适感角度上来讲，从室外自然环境阳光充足明媚的情况下，突然进入内部的蓝色环境，在整体上会给人不舒服的体验。

（2）从方法角度上来讲，该方案的设计已经达到了初步认知并联系运用装配化的程度。

（3）"6+"的活动强调了主题的重要性，是要能够解决实际存在的问题。

（4）从整体表达的角度来讲，逻辑性和图纸量比中期更完整。装配化的最终目的其实是多样化，它既要符合装修类型，也要与美感、舒适和审美的标准互相匹配。

The homestay design in the scheme of this group is featured by a unique style and is with full consideration. However, it is important to pay attention to how to explore the combination of the green and assembly color, highlight the characteristics of each other and coordinating the relationship between them. As far as the whole system is concerned, especially the spatial composition of this block shall be taken into account and applied to the overall context of the spatial construction. For example, the shades, walls, ceilings and the local scale of the home just mentioned and the digital representation are rational and rigorous. The content of these parts is needed, but we also want it to be reflected in the large space structure. Although the designer's consideration may not be so comprehensive, in the later period, we can make more in-depth supplementary explanations to improve the content of the scheme. Personally, a step-by-step and logical content framework would be better. Because the best content logic is to transition from the big aspect to the small aspect, from the whole to the local, from the shallow to the deep rather than highlighting the specific size of the details at the beginning, which presents a sense of the absence of an intermediate system.

The thematic nature of the scheme of this group is more and more topical than that in the medium term. There is a preliminary understanding and grasp of the concept of assembly. But why the ground has to be overhead, there is no clear reason for the proposal. For example, thickening the wall board is to strengthen the insulation of the wall; overhead floor is to ensure heating and ceiling is to solve the layout of air conditioner and fresh air system. So the assembly, as a technical means, is still a little stiff in the design.

(1) From the perspective of the beauty and comfort, suddenly entering the internal blue environment in the outdoor natural environment with sufficient sunshine will present people an uncomfortable experience on the whole.

(2) From the perspective of method, the design of this scheme has reached the level of preliminary cognition and application assembly.

(3) The" 6+" campaign emphasizes the importance of the theme to be able to solve practical problems.

(4) From the perspective of the overall expression, logicality and drawing quantity are more complete than the mid-term. The final goal of the assemble actually is diversification. It shall be in line with decoration types already, also match each other with aesthetic feeling, comfortable and aesthetic standard.

第二届中国国际进口博览会国家馆展示设计

National Art Exhibition Design of the 2nd China International Import Expo

高　　校：南京艺术学院

College: Nanjing University of the Arts

学　　生：王紫荆、王紫薇、朱一丰、高榕泽

Students: Wang Zijing, Wang Ziwei, Zhu Yifeng, Gao Rongze

指导教师：朱飞

Instructors: Zhu Fei

课题评价：85

Subject Scores: 85

王紫荆

王紫薇

朱一丰

高榕泽

学生感悟

Student's Thought

通过"6+"这个比赛，让我跟其他院校的同学学到了面对不同问题应采取不同的理解方式和解决方法。让我有机会与其他院校的同学进行专业上的交流，感谢"6+"和老师给我提供这么好的平台。

在学习过程中，我发现装备化即是限制又是优势，可以利用装备化的整体性统一展示国家形象并且形成非常具有带入感的视觉效果，这将成为我以后设计的宝贵经验之一。

读万卷书行万里路，经历越多感受越多，"6+"的联合毕业设计让我学到很多，也让我体会到自身还有的不足，这一活动充实了我学习设计的经历。

参与这次"6+"活动，让我从不同的专业学到的不同的知识，也有了新的看待设计的角度。

冰岛馆展示设计

Exhibition Design of Iceland Pavilion

王紫荆　王紫薇

展会定位 ┃ The Exhibition Location

第二届中国国际进口博览会国家馆展示要求

定位：面向专业观众展示新兴技术产业，面对普通观众展现国家形象；国家馆作为第二届中国国际进口博览会主要对外宣传的区域，是普通观众了解该博览会的主要区域。

绿色展位"6R"概念

对绿色展位、绿色运营、绿色物流、绿色餐饮的多项具体指标做了明确规定。绿色展位需要全过程遵循"6R概念"。

元素提取 ┃ Element Extraction

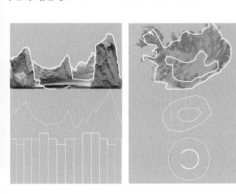

将冰川的造型抽象化，进行元素提取，用以体现展墙的高低起伏。

结合冰岛"碗状高地"的地形特点，组合成冰岛馆的围合方式。

冰岛丰富的地热资源为其提供清洁能源，冰与熔岩火山的巨大反差使冰岛拥有全球独一无二的自然风光。

空间功能与流线 ┃ Spatial Function and Streamline

功能与流线关系

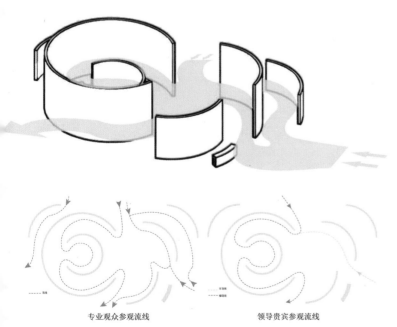

专业观众参观流线　　　　　领导贵宾参观流线

观众分析

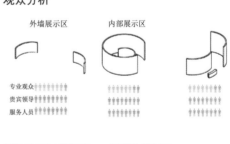

外墙展示区　　　内部展示区

专业观众
贵宾领导
服务人员

宣传视频展示时长：3min
区域步行距离：25m
区域参观时间：5～8min

展品数量：12件
区域步行距离：12m
区域参观时间：3～5min

冰岛馆预计每小时参观人次：16～327人次
预计参观时长：专业观众：5～15min
领导贵宾：8～13min

空间布局 丨 Space Layout

通过对空间形式、人的行为、文化等多角度之间的关系的表现，将冰岛人民与自然的相处模式融入空间当中。该案例主要体现冰岛在地热能源上的利用并发展旅游业和工业。展馆内将冰川地貌和其中蕴含的巨大能量用装配化的手段展现出来，内部展厅视同无纸化线上宣传表达冰岛对零碳上的坚持。从展厅外部看是冰的纯净，到展厅内部是颜色鲜艳的能量流动，强烈的对比也展现了冰岛人民与自然之间深刻的联系，同时也反映了冰岛人民对纯能源利用的智慧。

内容布局

我们将冰岛人民对地热的四个不同阶段的运用融入到我们的展厅平面布置之中。突出冰岛的纯净，将地热能作为贯穿整个展厅的内容。

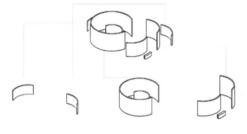

爆炸图

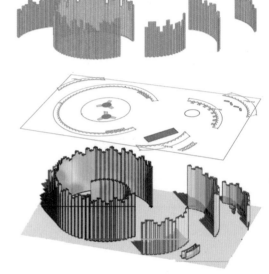

装置分布

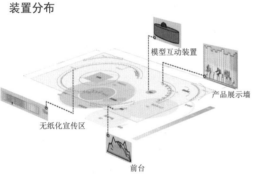

模型互动装置

产品展示墙

无纸化宣传区

前台

参观者停留点　　　　　流线热点　　　　　区域热度　　　　　区域划分

展墙内容

冰岛企业产品展示　　　　　地热能源应用技术介绍和案例展示　　　　　旅游资源和医疗设备展示

  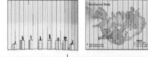

『室内设计 6+』2019（第七届）联合毕业设计
"Interior Design 6+"2019(Seventh Year) Joint Graduation Project Event

互动装置 | Interactive Device

线上展示

　　为了达到在短时间内让参观者获取大量信息，并能让参观者留下深刻印象，本案例秉持绿色环保的概念，宣传纸张全部由线上展示代替，使用 VR 设备和通过扫描产品旁边二维码的方式，不仅增加参观者的互动积极性，也迎合了现代成年人的工作方式，将展品资料通过微信等 APP 保留下来，也为双方合作提供了更多的可能性。

　　二维码：使用微信扫描获取。

　　VR 设备：头戴式，观看展厅全景或冰岛著名景点全景。

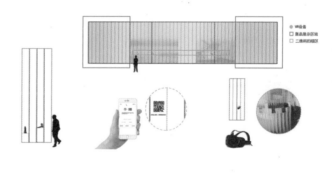

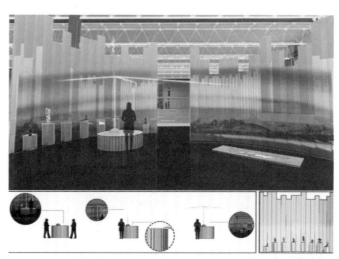

　　能量感应装置，目的是让参观者感受到身边无处不在能量。

　　参观者走近该装置时，靠近参观者的能量装置将红色的一面缓缓转向参观者。鲜艳的红色代表着能量，黑色的镜面玻璃代表大地。

　　激光感应装置，目的是为了呼吁使用清洁能源，减少碳排放，还原天空的纯净。

　　多名参观者通过触碰到感应桌，从地热发电厂模型的顶部射出激光，激光在遇到折射球后，四散在周围的展墙上，展墙上显示出美丽的北极光。

装配结构 I Assembly Structure

　　展墙以长为1200mm的玻璃板为基础件，所有展墙都可通过叠加和拼装组合。玻璃板的卡槽长度同样也为1200mm，并以此为基础单位，可以进行延长和拼接。
　　单元件通过霍克公司提供的镶嵌卡槽技术连接。

展墙尺寸高度

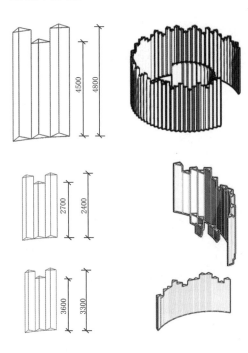

展具元素提取

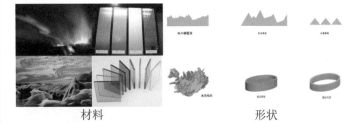

材料　　　　　　　　　　形状

玻璃展墙和玻璃显示屏

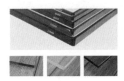　　

钢化玻璃　　　　中空玻璃　　玻璃显示屏：OLED 玻璃

展架拼装示意图

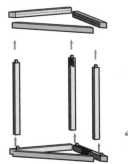 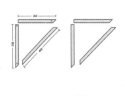 

霍克公司无孔玻璃爪装备示意图

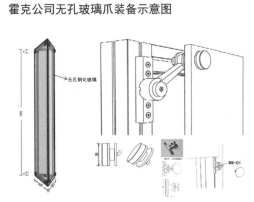

无孔钢化玻璃

展馆展具所用数量

展具规格/mm	450	750	展架尺寸/mm	150	300	1200	玻璃尺寸/mm	450×400	450×282	750×400	750×282	282×141	1200×400	1200×282
数量/个	12	20	数量/个	32	252	412	数量/个	16	32	26	52	392	331	662

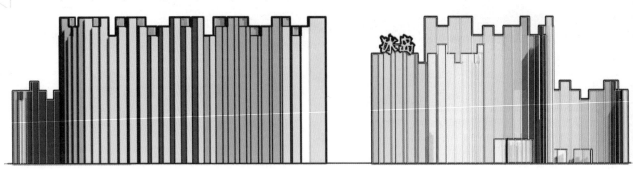

『室内设计 6+』2019（第七届）联合毕业设计 · "Interior Design 6+"2019(Seventh Year) Joint Graduation Project Event

中国馆展示设计

Exhibition Design of China Pavilion

朱一丰　高榕泽

概念生成 ┃ Concept Generation

我们分析研究了中国从古至今的对外展示形象，寻找一种可以表现当下中国综合形象的理念。结合博览会所需的商业模式以及装配化设计的要求，从新的角度看待中国，这是一个不断发展，传统与现代和谐并存的发展中国家。

平面生成 ┃ Plane Generation

天圆地方

"三进"式

亭，民所安定也

流线分析 ┃ Streamline Analysis

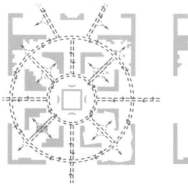

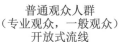
普通观众人群
（专业观众，一般观众）
开放式流线

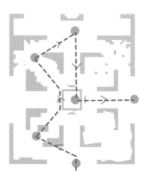
特殊观众人群
（专业观众，一般观众）
领导式流线

灯光变化 ｜ Lighting Change

中国馆的展墙框架是由统一的装配化结构搭建的，而结构中内嵌的 LED 灯带可以在调控下展示出多种主题的灯光变化，如快速生长的竹子、万家灯火般星星点点的光芒、园林中的中式格栅样式等。在无灯光展示的情况下，白色框架又如同一座座快速搭建的钢铁森林，这让整个展示空间有了多重变化，不仅是一个展示空间同时也是一个艺术装置。

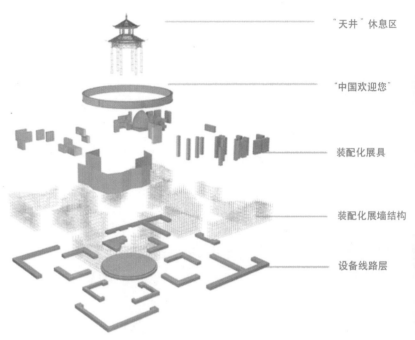

"天井"休息区

"中国欢迎您"

装配化展具

装配化展墙结构

设备线路层

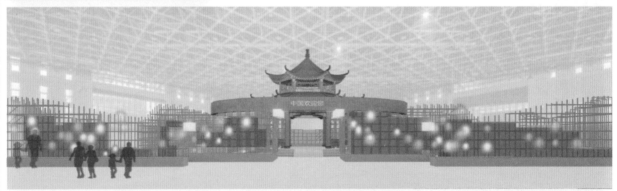

「室内设计 6+」2019（第七届）联合毕业设计
"Interior Design 6+"2019(Seventh Year) Joint Graduation Project Event

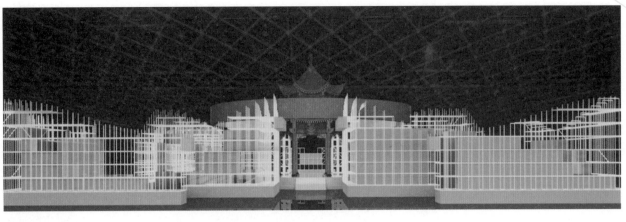

装配化思路 | Assembly Thinking

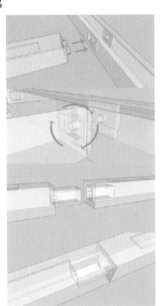

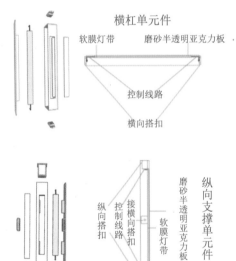

横杠单元件

软膜灯带　　　　磨砂半透明亚克力板

控制线路

横向搭扣

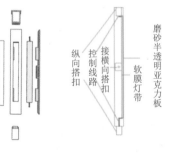

纵向支撑单元件

磨砂半透明亚克力板

接横向搭扣

控制线路

软膜灯带

纵向搭扣

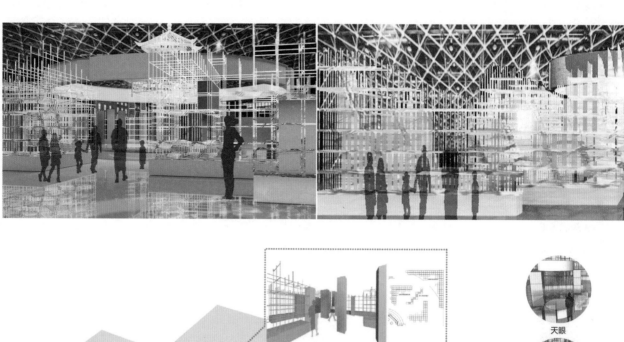

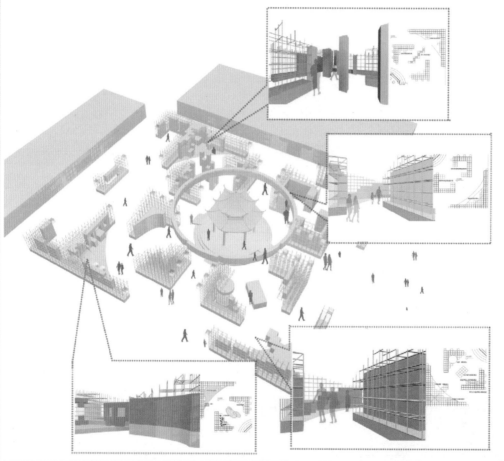

天眼

中欧班列

库布齐沙漠治理

港澳台大桥

复兴号高铁

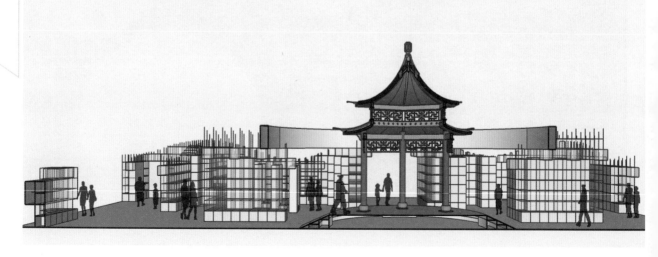

模型效果图

模型展览图

徐敏

这一组同学学的是展示专业，正好又选了展示的题目。在整个设计的过程中，反应表现相当不错。总体来讲，有两方面优点：①分析思考的过程具有严格的逻辑性。在国家的选择上，选择了具有差异化较大的冰岛和中国。这两个国家的文化底蕴或文化形象是截然不同的。②在装配化方面，由于学生本身的专业性的较强，对于展述和装配性分析得很好。

不足之处是，对国家形象的塑造有欠缺。因为这个主题是个老话题，世界博览会中的中国馆在很长的一段时间饱受争议。可能大厅的信号不好，关于绿色进口博览会是什么主题内容，我没有接收到。

刘杰

前期详细的调研分析支撑了后续的具体设计和实施，例如参观完整个展厅大概需要花费多少时间，这项工作做得很好很细致。另外尽管有展位的模型和比例尺，还是希望在图纸上能看到立面尺寸标注，这样对展位和整个展馆的高度和比例有个更好的了解。另外中国馆亭子的创意，营造出强烈的传统符号和语境，可以将构建语言做得简洁一些，更符合我们时代对于中国元素符号的理解，并且大众更容易接受。

洪亚妮

冰岛馆所传达的概念留给本人的印象非常深刻，该方案设计的切入点很独特，阐述清晰，整个展示空间的形态富有张力。本人刚从冰岛回来，方案的概念设想和本人脑海中的认知是有契合度的。

很高兴听到方案中有装配模块和构成逻辑的阐述，但也发现了以下问题：

（1）构建结构表述得不够清楚，三角交接的关系存在漏洞，玻璃界面的接口形式不够清楚，光影等细节还需推演，该方案可以再深入细化一些。

（2）方案中展示空间的形态就其本身也是展示内容之一，这里建议要补充其他的展示路径设置，构建更多具象展示内容的表达方式，并将这些方式交融在整体展示状态中。

总体而言，概念切入点非常好，但落地性还在路上，期待这个毕设节点也是大家下一个设计旅程的开始。

［室内设计 6+］2019（第七届）联合毕业设计
"Interior Design 6+"2019(Seventh Year) Joint Graduation Project Event

Students of this group major in display and chose the topic of display. In the whole design process, they made quite good response. From several ways, the advantages are shown as follows: 1. The process of analyzing and thinking is strictly logical. In the choice of the country, Iceland and China with different contrasts were chosen. The cultural heritage or cultural image of these two countries is quite different. 2. In terms of assembly, due to its strong professionalism, they conducted presentation and assembly analysis well.

Disadvantages: 1. It affects the shaping of national image. Because the theme is an old topic, the China pavilion at the Expo has been controversial for a long time. Maybe the signal in the hall is not good. As for the theme of Green Import Expo, I have not received it.

The detailed investigation and analysis in the early stage supported the subsequent specific design and implementation. For example, how much time it would take to visit the whole exhibition hall. This work was conducted very well and carefully. In addition, although there is a model and scale of the booth, we hope to see the dimension mark of the facade on the drawing, so that we can have a good grasp of its height matching with the whole exhibition hall. And the creativity of the Chinese pavilion pavilion creates a strong traditional symbol and context, which can make the construction language concise, more in line with our understanding of the Chinese symbol in our time, and more easily accepted by the public.

The Icelandic pavilion concept is very impressive. The design of the scheme is unique, with clear description; and the shape of the entire exhibition space is full of tension. I have just returned from Iceland, and the concept of the scheme is in line with the cognition in my mind.

I am very glad to hear the explanation of assembly module and composition logic in the scheme, but I also have found problems:

(1) The construction structure is not clearly expressed, there are loopholes in the triangular connection, the interface form of the glass interface is not clear enough, and details such as light and shadow, etc. still need to be deduced, so the scheme can be further refined.

(2) Another problem is that the form of display space in the scheme is one of the display contents itself. It is suggested to supplement other display path settings, build more expressive ways of display contents, and integrate these ways into the overall display state.

In general, the conceptual entry point is very good, but it is still under implementation. We expect this graduation node to be the beginning of your next design journey.

答辩评审

高　　校：浙江工业大学

College: Zhejiang Unversity of Technology

学　　生：范诗意、陈格

Students: Fan Shiyi, Chen Ge

指导教师：吕勤智、王一涵

Instructors: Lü Qinzhi, Wang Yihan

课题分数：90

Subject Scores: 90

基于一体设计与装配化建造的阿里巴巴青年公寓室内设计研究

Interior Design of Alibaba youth Apartment Based on Integrated Design and Assembly

范诗意

陈格

学生感悟

Student's Thought

　　从最初的了解选题内容、文献查阅，到之后的实地调研、人群采访，再到最后设计框架以及整个方案的呈现，整个过程非常的辛苦，但同时又有许多的收获，并且和搭档一起很好地完成了大学最后的项目。非常感谢老师的指导以及搭档的配合。

　　通过这次毕业设计，不仅对自己大学四年所学的知识有所检验，同时也提高了自己的设计能力。感谢两位指导老师这段时间耐心且细致的指导，让初次接触选题还有些糊涂的我逐渐找到了自己的方向。

轨道交通系统 | Rail Transit System

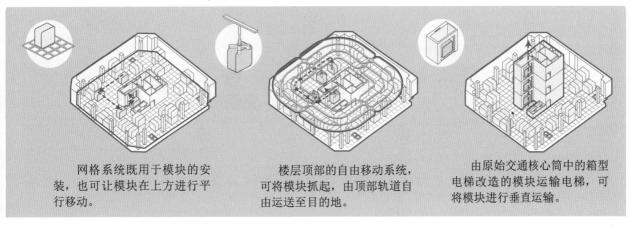

网格系统既用于模块的安装，也可让模块在上方进行平行移动。

楼层顶部的自由移动系统，可将模块抓起，由顶部轨道自由运送至目的地。

由原始交通核心筒中的箱型电梯改造的模块运输电梯，可将模块进行垂直运输。

功能模块菜单式选择 ┃ Rail Transit System

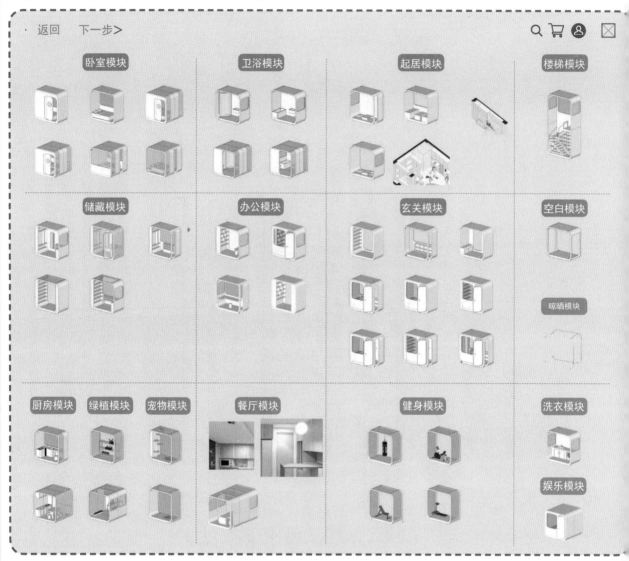

模块组合类型 ┃ Module Combination Type

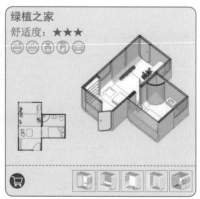

　　用户通过对各个功能模块的菜单式选择，进而组成一个属于他们自己的家。我们也对家的组合类型进行了一定的归纳，住户可参考我们所归纳的类型，也可按照自己的意向进行选择拼装，选择完毕后系统也会给出合理的交通组织方式。

『室内设计 6+』2019（第七届）联合毕业设计
"Interior Design 6+"2019(Seventh Year) Joint Graduation Project Event

宠物之家
舒适度：★★★☆

健身之家
舒适度：
★★★★

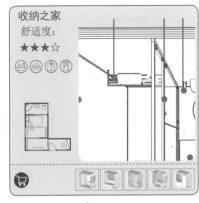

收纳之家
舒适度：
★★★☆

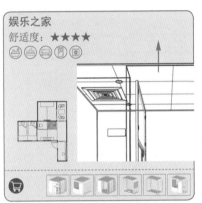

娱乐之家
舒适度：★★★★

健身双人之家
舒适度：★★★★

相对隐私的双人之家
舒适度：★★★★

标准双人之家
舒适度：★★★★☆

1F　2F

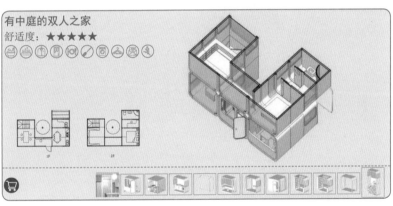

有中庭的双人之家
舒适度：★★★★★

1F　2F

答辩评审

互联网居住系统 App ｜ Internet Living System
"家"的选择

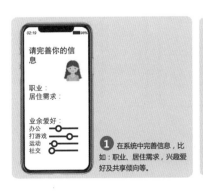

请完善你的信息

职业：
居住需求：

业余爱好：
办公
打游戏
运动
社交

① 在系统中完善信息，比如：职业、居住需求、兴趣爱好及共享倾向等。

请选择你想居住的位置及楼层

请选择你所喜欢的空间模块

卧室模块　办公模块
玄关模块　储藏空间模块

请选择你所喜欢的材质

● Wood

请选择家具模块

② 根据系统操作，选择居住位置、空间模块、家具模块、材质颜色等，通过VR进行模拟拼装。

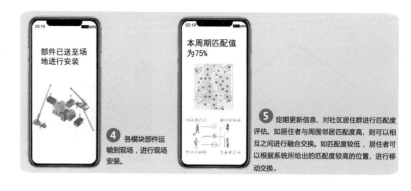

③ 订单完成，工厂制作。

④ 各模块部件运输到现场，进行现场安装。

⑤ 定期更新信息，对社区居住群进行匹配度评估。如居住者与周围邻居匹配度高，则可以相互之间进行融合交换。如匹配度较低，居住者可以根据系统所给出的匹配度较高的位置，进行移动交换。

社区共享系统

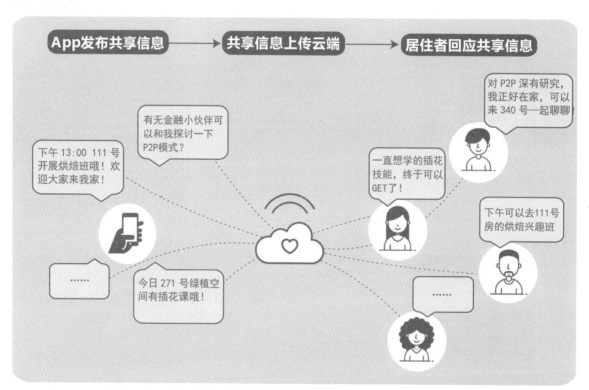

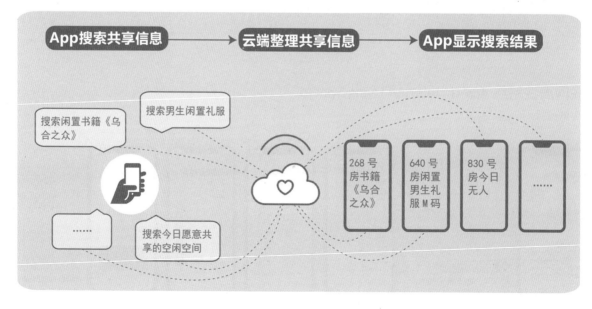

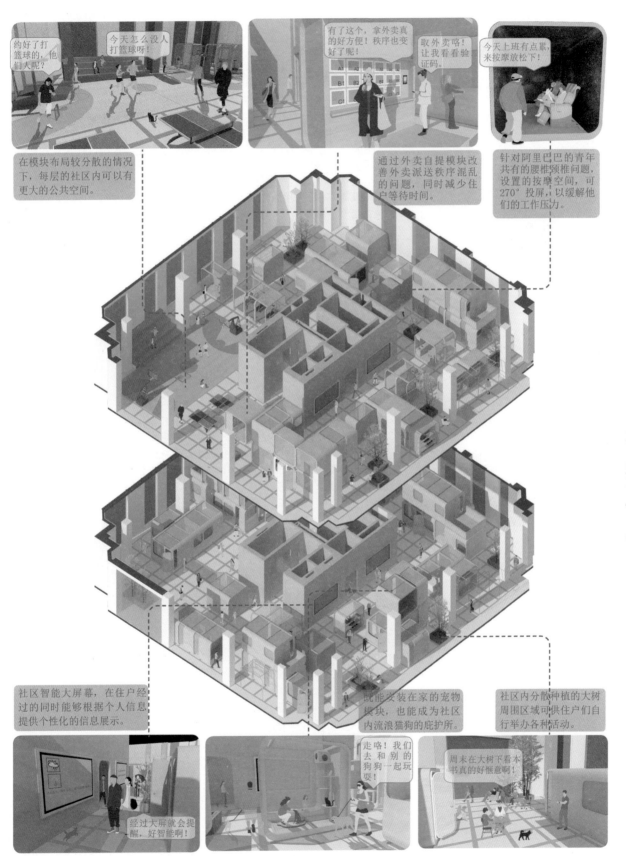

姚渊明

集成化可以说是自始至终贯穿整个设计中，扣题很好。这是设计方案必须需要的温度，在家具、软装和艺术品各方面均有体现。通过前面 13 组同学的汇报，本人发现提及照明设计的内容很少。好的设计都是去繁成简，同学们要提高总结归纳的能力。同学们即将步入职场，传统的相关理论知识概念也非常重要，例如最基础的中外建筑史等都是十分有用的。

谢冠一

这些单元组合也许不成问题，但组合的效果往往是让人担忧的，因为太像车厢的形态了。考虑连接的效果通常是单方向的，而不是多方向的。究竟是垂直还是横向构造，肯定会碰到技术性的难点。该组方案让我们看到了另一种努力的方向，其以使用者的角度来切入表达，并不是关注于单元模块之间组合出来的空间效果，而是更多地去陈述怎样组成系统，这种系统会带来什么样的生活方式，可以解决什么样的生活困难。可以说，以后的职业发展不一定就局限于专业设计师，当然也能转型到周边的行业。我想这应该是受到老师的启发多一些。事实上，单元组合的秩序、节奏和意义更加有待考虑。我们要以包容心来对待这种目前可能不是太成熟的设计思考方向，我相信这会带来新的生活形态和新的生活方式。

Integration can be said to run throughout the whole design, with extremely good points, which is very important for the design scheme; and integration is reflected in all aspects of furniture, soft decoration and artwork. Through the report of students of the first 13 groups, I have found that it mentioned less about lighting design. The good design is to covert complex to simple and improve the ability to summarize. Students are about to enter the workplace and traditional theoretical knowledge and concepts are also very important. For example, the most basic eastern and western architectural history, etc., which are very useful.

The combination of these units may not be a problem, but we concern the effect. Because it's too much like a carriage, and the effect of the connection is usually unidirectional, not multi-directional. Whether it is vertical or horizontal structure, technical difficulties must be encountered. This scheme of group makes us see another direction of efforts, which focuses on the expression from the perspective of users, instead of the spatial effect combined between modules, but it pays more attention to that how to form a system, what kind of lifestyle this system will bring and what kind of life difficulties it can solve. It can be said it is not necessary to limit the future career development to professional designers, but it can also transform to surrounding industries. I think this shall be inspired by the teacher more. In fact, the order, rhythm and meaning of the unit combinations are more to be considered. We shall treat this design thinking direction, which may not be mature at present, with tolerance. I believe it will bring new life forms and new lifestyles.

热点命题　纷显特色　　联合指导　服务需求

教育研究

「室内设计6+」2019（第七届）联合毕业设计

"Interior Design 6+"2019(Seventh Year) Joint
Graduation Project Event

智园行方

百年大计 立德树人 专业造就 质量第一
以创新能力培养为主线 全面系统地总结教
育教学理论与实践

小户型租赁住宅的春天——深圳城中村住宅改造设计

左琰　林怡

Spring of Small Rental Housing—Housing Renovation Design of Urban Villages in Shenzhen

Zuo Yan, Lin Yi

2019 中国室内"6+"联合毕业设计的总课题为"绿色营造——建筑室内装配化设计"，同济大学此次与上海全筑建筑装饰集团股份有限公司合作，聚焦深圳城中村住宅改造，以高密度状态下的小户型租赁住宅为设计对象，探讨其如何实现私密性、舒适性和灵活性，以及内部装饰与空间、家具、部品、机电等多个领域整合的可能性。

白石洲是深圳最大的农民房集中区，建筑密度高、容量大，环境复杂。刚刚进入社会的青年群体俨然成为深圳等特大型城市新一代的弱势群体，他们面临着巨大的生存和社会压力，困扰他们的最大问题是住房。对于人口拥挤的一线城市，高密度复合型的单元模式已经成为有效的解决方案。以小面积为出发点的小户型公寓及住宅应运而生，并日益成为青年人喜爱的一种房型。

学生 8 人分成两组，在前期展开基地调研并对北京、上海和深圳等城市租赁住宅市场进行分析研究，对租赁对象的年龄、职业、生活方式和家庭构成等因素进一步分类后形成对应的户型平面空间，因为是小组合作，对 37 号住宅楼的改造包括了建筑外观立面、公共空间和户型平面室内设计、家具模块化设计等多个方面。"晴天见"小组对租赁对象的性别有特别关注，推出了女性户型和 10m² 左右带卫生间的极小样板，而"90后的理想家"注重外观立面的改造，户型平面细分成单身或情侣小户型以及三口或三代同堂等偏大户型，并对卫生间展开了装配化体系设计。两组学生通过在建筑室内一体化、室内部品一体化、室内光环境设计与整合等多方面来改善和提高租户的生活品质。

建筑立面改造一方面是建筑形象设计，也是建筑外围护体如何处理室内与外部环境物理关系的问题。由于白石洲现有建筑密度极高，楼宇与楼宇间的超近距离，导致私密性较差，室内空间的采光受到影响。因此立面改造强调采光效果与保证室内私密性的平衡。要求学生对基地外部条件、室内功能需求进行针对性分析和采光条件研究，提出相应的设计策略和技术手段。两个小组的设计都在对其立面功能属性需求分析的基础上做出了较好的设计回应。"90后的理想家"采用立面竖向遮阳设计，在阻隔直射眩光、保障私密性的基础上增加了室内光照的均匀性，提升了室内光环境品质，并根据对光环境的需求，设置居住空间的不同功能区块的平面位置，在小户型空间中实现功能最大化

The total proposition of "The Interior Design 6+" Joint Graduation Design 2019 is "Green construction – building indoor assembly design". Tongji University cooperated with Shanghai Trendzone Villa Deco Group Co., Ltd. this time to focus on the housing renovation of urban villages in Shenzhen, adopt the small rental housing in high density as the design object, discuss the way to achieve its privacy, comfortableness and flexibility, and the probability of the integration in various fields of interior decoration, space, furniture, components and parts, and electromechanical equipment, etc.

Baishizhou is the largest farmer housing concentration district in Shenzhen with high building density, large volume and complicated environment. The youth group that has just entered the society has become the disadvantaged group in new generation in the metropolis including Shenzhen. They face great survival and social pressure, and the greatest problem of housing that bothers them. In the highly-populated highly-developed cities, the model of highly dense and compound unit has been an effective solution. The small apartment and residence focusing on small area comes into being at the right moment, and has gradually become the house type popular among the youth in current urban real estate market.

Eight students, divided into two groups, preliminarily carried out base survey, analyzed and researched the urban rental housing market in Beijing, Shanghai, and Shenzhen, etc., further classified the criteria including age, occupation, lifestyle and family composition of the rental objects, etc., and formulated the floor plan of corresponding house type. With team cooperation, the renovation of residence building No.37 included building external façade, pubic space, interior design of house type plan, furniture modular design, etc. The team called "See you in sunny day" paid special attention to the gender of rental objects, and released female house type and mini housing model of about 10m² with toilet. The team called "Post-90s ideal home" paid attention to the renovation of external façade, further divided the house type plan into small house type for single person or couples, and relatively large house type for family of three or family of three generations, and undertook assembly system design. The two groups of students improved and enhanced the living quality of the tenants in multiple fields of building interior design integration, indoor component and part integration, indoor luminous environment design and integration, etc.

The renovation of building façade not only covers building image design, but also relates to the issue how to deal with the physical relation between indoor and external environment through building external protector. The extremely high density of the existing buildings, limited natural lighting of the buildings, and too short distance between buildings in Baishizhou aggravate the privacy issue and the obtaining of natural light in existing indoor space. Therefore, façade renovation emphasized the balance daylighting effect and the guarantee of indoor privacy. The students were required to accordingly analyze the external conditions of the base and indoor function demands and research the daylighting conditions, and propose corresponding design strategy and technical means. The design of the two groups relatively well corresponded in the design based on the analysis of façade function and property demand. "Post-

满足。"晴天见"小组则采用了完全不同的设计手法，利用水平楼板出挑，结合异形阳台设计，实现了建筑构件的自遮阳，并通过阳台形成与外界环境的过渡空间，一定程度上保障了室内活动的私密性。

90s ideal home" adopted façade vertical sunshade design, increased the evenness of indoor lighting on the basis of obstructing direct glare and protecting privacy, improved indoor lighting environment quality, and set the plan position of different functional sections of the living space according to luminous environment demand, and realized the functions in small house type space to the greatest extent. "See you in sunny day" team achieved self-sunshading of building component with horizontal floorslab protrusion in different design method, and protected the privacy of indoor activities to certain extent with the transitional space in the balcony to the external environment.

教育研究

"室内设计 6+"联合毕设实践活动精准扶贫模式初探

杨琳　张博涵　马宇萌

一、我们都是追梦人

春暖花开，第七届"室内设计 6+"联合毕设开题活动在北京建筑大学举办，来自 7 所高校、5 家企业、3 家媒体的 20 多位导师和 40 多名学生参加了开题汇报、论坛及考察活动。

本届联合毕设以"绿色营造——建筑室内装配化设计"为总命题，探索装配式建筑、发展绿色建筑、讨论先进建造方式，为建筑与室内装配化设计专门人才培养做出积极的推动。

开题活动中，企业导师刘恒院长分享的"建设可再生的未来——雄安设计中心环境设计、雄安实地调研"，使在场的师生了解了绿色营造、设计总承包、建筑师负责制等一系列绿色营造理念在实际项目中的应用。

在开题活动最后，由哈尔滨工业大学和北京建筑大学组建的内蒙古阿尔山市明水河镇西口村民宿示范区环境设计团队，将为乡村振兴提供设计服务。

二、踏勘阿尔山西口村

两会期间习近平总书记参加内蒙古代表团审议时强调，保持加强生态文明建设的战略定力，守护好祖国北疆这道亮丽风景线；保护草原、森林是内蒙古生态系统保护的首要任务；保护生态环境和发展经济从根本上讲是有机统一、相辅相成的；打好脱贫攻坚战，不断增强人民群众获得感、幸福感、安全感。

值此时机，参与"室内设计 6+"联合毕业设计的哈尔滨工业大学以及北京建筑大学师生赴西口村进行实地踏勘。西口村民宿示范区位于世界地质公园阿尔山景区南出口沿线的明水河镇，占地 9000m²，周边绿水青山，是绝佳的休憩目的地。当地因每年举办的"赶大集"活动，在周边方圆几十里颇具盛名，具有丰富的历史人文气息和蒙元文化特色。由于林场转型，西口村 1000 多户村民中有 350 户建档立卡户，是脱贫攻坚的重点村。

哈尔滨工业大学及北京建筑大学师生对村容村貌、民居建筑、当地生活习俗、扶贫成果等进行了参观，并实地测量了示范区场地及建筑，与村委会干部、建档立卡贫困户座谈，收集并完善了设计工作所需的基础资料。

三、助力乡村文旅融合发展

内蒙古阿尔山市是文化和旅游部（简称"文旅部"）精准扶贫地区。阿尔山市下辖的西口村地处阿尔山全域旅游区域南部交通线上，将持续打造文旅融合的乡村旅游目的地。"室内设计 6+"联合毕业设计团队师生与文旅部挂职驻村第一书记李绘新等村委干部公用研讨、策划，初步

I. We Are All Dreamers

The seventh "Interior design 6+" joint graduation opening activity was held in Beijing University of Civil Engineering and Architecture. More than 20 tutors and 40 students from 7 universities, 5 enterprises and 3 media participated attended the opening report, forum and investigation.

The general theme of this year's "green building—prefabricated interior design" is to explore the development of prefabricated buildings, green buildings and advanced construction methods, and actively explore the development of prefabricated architecture and interior design experts.

"Building a renewable future—Xiongan design center's environmental design and Xiongan's field research", enable teachers and students to experience a series of contents such as green construction, general contracting of design and responsibility system of architects.

In the opening activity, Harbin Institute of Technology and Beijing University of Civil Engineering and Architecture established the environmental design team of Xikou villager's residence demonstration area, Mingshuihe town, Arxan city, Inner Mongolia, to serve the rural revitalization.

II.Visiting Xikou Village of Arxan

During the NPC and CPPCC sessions, general secretary Xi Jinping attended the deliberations of the Inner Mongolia delegation, stressing that we should maintain our strategic focus on strengthening ecological civilization and protect the beautiful scenery of northern Xinjiang. Protecting grassland and forest is the primary task of ecosystem protection in Inner Mongolia. Protecting the ecological environment and developing the economy are fundamentally integrated and mutually reinforcing. We will overcoming poverty and continue to increase the sense of gain, happiness and security among our people.

Using this opportunity, teachers and students of Harbin Institute of Technology and Beijing University of Civil Engineering and Architecture, who participated in the joint graduation design of "6+ interior design", went to Xikou village for field survey.

Located in Mingshuihe town along the south exit of Arxan Global geopark, Xikou villager's residence demonstration area covers an area of 9,000m², surrounded by green water and mountains, and is an excellent rest destination. It is well known for its grand village market in hundreds of square miles, with rich historical and cultural atmosphere and Mongolian yuan cultural characteristics. Due to the transformation of forest farm, Xikou village has 350 registered households among more than 1,000 households, which is the key village for poverty alleviation.

The teachers and students of the "interior design" studio surveyed the appearance of the village, residential buildings, local customs and poverty alleviation achievements exhibition, measured the site and buildings of the demonstration area, discussed with the cadres of the village committee and the poor households with established files, collected and improved basic design materials.

III. Help the Integrated Development of Rural Culture and Tourism

Arxan city, Inner Mongolia, is aiming at poverty alleviation by the ministry of culture and tourism. Xikou village, located on the southern transportation line of Arxan regional tourism region, will continue to build a rural tourism destination integrated with cultural

确定了西口村民宿示范区品牌化经营策略。

通过乡村文化环境设计，将示范区公共环境塑造为展现蒙元文化、乡村风情、体验传统文化、非遗工坊的场所，满足展览、展示、展演、展销体验等功能需求。民宿则保留传统火炕火墙等民俗特色。同时，改建现代种植温室，为冬季旅友提供农事体验等一系列内容，力求打造青山绿水白雪生态村。并运用绿色建造技术和部品部件生产工厂化、施工装配化设计，在改造工程中减少对环境的影响，与自然和谐共处。

四、探索扶贫扶智模式

在"室内设计6+"多元培养环境提出的综合性实践教学活动，通过文旅部和校企导师、师生的通力合作，取得了西口村民宿示范区环境设计成果。以环境设计为载体，内蒙古阿尔山民宿品牌化经营模式雏形初现。今后希望能够带动更多西口村民和阿尔山市其他乡镇村民参与文旅服务，争取增收脱贫并能达到经济的持续增长。

同时，"室内设计6+"团队与西口村建立的民宿设计平台，也将持续针对村民不同类型的宅院改造需求提供设计咨询服务。

7月，习近平总书记再次赴内蒙古，就经济社会发展、生态文明建设进行考察调研，实地指导开展"不忘初心、牢记使命"主题教育。在未来，通过当地政府、村民以及"室内设计6+"设计团队的共同努力，将把祖国北部边疆这道风景线打造得更加亮丽。

tourism. "Interior design 6+", organized by teachers and students of the graduation design team and Li Huaxin, the chief secretary of the ministry of culture and tourism in the village, and other village committee cadres to discuss and plan the brand management strategy of Xikou villagers'lodging demonstration area.

Through the design of rural cultural environment, the public environment of the demonstration area is shaped into a place to display Mongolian yuan culture and rural customs, experience traditional culture and intangible cultural heritage workshop, and meet the functional requirements of exhibition, performance, sales. A series of measures, such as retaining the folk characteristics of traditional heating beds and heating walls, rebuilding the modern planting greenhouse to provide agricultural experience for winter travelers, and striving to build an ecological village of green mountains, water and snow. And the use of green construction technology and parts of the production of factory, construction assembly design, reduce the impact on the environment, and harmony with nature.

IV. Explore the Model of Poverty Alleviation and Intellectual Support

In "interior design + 6" pluralistic culture of comprehensive practice teaching activities, the brigade headquarters and between teachers, tutors and students work together, the characteristics of the villagers lodge demonstration area environment design, environmental design as the carrier, stretching over home stay facility brand management pattern is taking shape in Inner Mongolia, hoping to bring more characteristics of the villagers and the city's rural villagers to participate in travel service, increase of poverty and sustained growth.

Meanwhile, the "interior design 6+" team established a homestay design platform with Xikou village to continuously provide design consultation and interaction for villagers' different types of house renovation.

In July, General Secretary Xi Jinping went to Inner Mongolia again to conduct investigation on economic and social development and ecological civilization construction, and to guide the education on the theme of " Remain true to our original aspiration and keep our mission firmly in mind ". In the future, through the joint efforts of local governments, villagers and the "Interior Design 6+" design team, the landscape of the northern frontier of the motherland will be more beautiful.

Discussion and Application of Fabricated Technology in Specialized Interior Design Courses

陈小丹

Chen Xiaodan

由于我国社会经济不断地发展，住宅产业化与健康绿色建筑的相关政策受到国家大力推进，室内设计行业中整体装修产业化将会迎来进一步的发展，与此同时相关技术与理念不断地更新，尤其是室内装配式装修将会迎来更大的发展机遇。装配式装修也称工业化装修，兴起于第二次世界大战后，20世纪在日本得到迅速的发展，从而逐步形成完整的装配体系。随着室内装配式技术研究的逐步开展和示范性工程的实践，现在室内设计教学中传统课程内容已不适合室内装修发展的要求。那么如何在传统内容的基础上，在课程教学中引入装配式技术理念，这是一个值得探讨的问题。在此基础上，本文主要从以下几点进行分析，并在传统室内设计的基础上进行改进与创新，以便为未来装配式建筑模式下的室内设计发展提供一定的参考。

一、装配式室内设计的基本概念

装配式装修是将工厂生产的产品、部件通过可靠的装配方式，由专业的产业工人按照标准化程序采用干法施工的装修过程。在装配式建筑模式中，室内设计不仅包括科学、合理、有效的具体设计工作，而且还是一个将建筑装饰设计从工程整体内容到策划、设计、生产、运输、安装等结合于一体的建筑装饰设计模式。

二、装配式装修的特点及市场需求

装配式装修体系的主要特点：设计标准化、生产工厂化、施工装配化（天花灯槽系统、墙面挂板系统、地面系统）、装修一体化（整体厨房系统、整体卫浴系统）、管理信息化。装配式装修与传统装修相比较而言，其现场只需产业工人能够严格执行施工工序，有效提高工作效率，减少现场施工周期；集成生产产品系统高度工业化，品质更稳定，更加节约原材料；现场组装，无灰尘污染与噪声，绿色环保。

社会发展的现阶段决定人们对今天装修品质的需求有所提升，因此装配式建筑和新建住宅的全装修的全面推进，是国家为促进建材产业转型升级和健康发展而采取的战略措施。装配式装修作为当前阶段室内装修发展的重要任务。不仅引发了室内装饰装修产业链的转型和服务的升级，同时也为企业的转型和升级带来了人才结构的变化，提出了新的技术需求。"产业转型，人才先行"，那么就要在高校中将装配式技术与室内设计专业课程相结合，面向室内装修当中设计标准化、生产工厂化、施工企业装配化，培养从事设计、生产和施工管理岗位群的高素质技术技能人才。

三、创新教学方法，激发学生的创新热情

1. 深入企业调研，构建课程体系

通过对装配式装修的设计、生产、部品集成企业的深入调研，总结、归纳、提炼相关专业技术的核心能力和知识结构，构建基础知识和基础技能的教学

Benefiting from the ongoing development of the Chinese social economy, policies on housing industrialization and healthy green architecture are strongly carried forward by the country. The overall industrialization of decoration in the interior design industry will further develop. Meanwhile relevant technologies and ideas keep updating, especially the interior fabricated decoration which will meet greater development opportunities. Fabricated decoration, also known as industrialized decoration, emerged after WWII and developed rapidly in Japan in the 20th century to gradually form a complete fabricated system. With the gradually launch of studies on interior fabricated technologies and practices of demonstration projects, the traditional courses in interior design teaching can no longer meet the requirements for the development of interior decoration, making it a discussion-worthy issue how to introduce the fabricated technical ideas to course teaching based on the traditional contents. On this basis, the paper mainly conducts analysis from the following points and improvement and innovation based on traditional interior design to provide certain reference for the development of interior design under the fabricated architectural mode in the future.

I.Basic Concept of Fabricated Interior Design

Fabricated decoration is a process of decoration in which dry construction is adopted by professional industrial workers according to the standard procedure to reliably fit the products and parts from factories. In a fabricated architectural mode, interior design not only includes scientific, rational and effective specific design, but also serves as a design mode of architectural decoration integrated from overall engineering to planning, design, production, transportation and installation, etc.

II. Characteristics and Market Demand of Fabricated Decoration

Main characteristics of fabricated decoration system: standardized design, factory-oriented production, fabricated construction (ceiling troffer system, wall hanging board system and ground system), integrated decoration (integrated kitchen system and integrated bathroom system) and information-oriented management. Compared to traditional decoration, fabricated decoration only requires industrial workers to strictly implement the procedures on site, which effectively improves work efficiency and reduces field construction cycle; the integrated production system is highly industrialized, providing more stable quality and saving more raw materials; the field assembly is free from dust, pollution and noise, making it green and environmental friendly.

The present stage of social development makes higher demand for decoration quality among people nowadays. Therefore it is the strategic measure of the country to comprehensively develop the full decoration of fabricated architecture and newly built housing to promote transformation, upgrade and healthy development of the building material industry. Fabricated decoration as a vital mission of interior decoration development in the present stage not only triggers the transformation of full industrial chains of interior decoration and the upgrade of services, but also brings change in talent structure to the transformation and upgrade of enterprises. As a new technical demand is put forward, "talent first for industrial transformation", the specialized courses in fabricated technology and interior design shall be combined in colleges and universities to cultivate high-quality technical talents on design, production and construction management positions for standardized design, factory-oriented production and fabricated construction enterprises in interior decoration.

III. Innovating Teaching Methods to Motivate Innovation among

体系，在课程建设过程中，通过加强校企合作，把装配式技术转化为教学资源和教材，同时将院校的课题研究成果向企业进行转化。在整个建设过程中，建立互利共赢的机制，发挥各方优势，激励校企协同发展与创新。

2. 实地现场教学，理论结合实践

传统的室内装修采用的施工工艺是工序复杂、耗时费力的湿作业，并且相关设计课程讲授主要是教导学生如何按部就班地进行设计施工。而装配式技术是一个创新的理念，因为学生在学习材料与施工工艺时就已经具备一定的基础知识，所以在课堂讲授理论的基础上我们可以将学生带到生产单位、施工作业单位，现场观摩装配式装修具体的施工工艺流程，这样不仅为学生在室内装饰的学习上开拓新的设计理念，同时也能够将理论知识的学习与实践相结合，让学生更好地运用于课程设计当中。

3. 优化课程设计，提升教学质量

装配式设计是系统和复杂的，所以学生在完成作业时可以采用小组协作的方式，同学们通过互相讨论对室内空间提出自己的装配式构造方案设计，再由代课教师和有装配式设计经验的设计师对其构造进一步指导，有针对性地解决造型设计和装配结构之间的联系，从而进一步完善方案，对于好的设计可以与厂家协作进行产品开发生产。室内设计课程的设置能够加强学生间的互动，发掘学生的潜力，组织学生就课程相关的学术前沿开展科技创新活动，针对不同教学内容以及不同学习阶段，采用讨论、示范、启发等教学方法，增强学生的学习主动性，引导学生自主而积极地参与学习，并进行探索创新，使学生真正成为学习的主体，鼓励学生积极思考，大胆实践，从而提高教学效果。

四、结语

当今装配式技术的应用不仅能提高室内装修的工作效率、工艺品质、环保性，同时也能体现现代化专业性设计。因此将装配式装修与室内设计专业课程相结合，不仅能够提高课程教学效果，同时对提升学生的课程设计能力起到关键性的作用，也能够让学生把书本上的理论性知识灵活地应用于实际的设计项目当中，在未来的设计中更具创新能力，也有助于培养工作高效率，设计高品质的优质大学生，同时推动我国绿色建筑和装配产业化室内建设的良好发展。

Students

1. In-depth Enterprise Research to Build Course System

Through in-depth research on design, production and parts integrated enterprises of fabricated decoration, the core competence and knowledge structure of relevant professional skills may be summarized, concluded and refined and the teaching system on basic knowledge and skills may be built. In the process of course construction, fabricated technology may be transformed to teaching resources and materials through reinforced cooperation between schools and enterprises. Meanwhile, the results of research projects in colleges and universities may be transferred to enterprises. In the whole construction process, a mechanism of mutual benefit and win-win may be built to perform their respective advantages and encourage synergetic development and innovation between schools and enterprises.

2. Field Teaching to Combine Theories with Practices

The construction technology adopted in traditional interior decoration is a complicated wet construction that consumes both time and energy. Moreover, the lecturing of relevant design courses mainly focuses on teaching student how to design and construct according to the routine procedure. In contrast, fabricated technology is a new idea. As students are equipped with certain basic knowledge upon learning materials and construction technologies, we can bring students to production units and construction units based on teaching of theories on classes so that they may view the specific construction process of fabricated decoration on site. In this case, students can not only explore new design areas in interior decoration learning, but also combine theoretical knowledge with practices for better application in course design.

3. Optimizing Course Design to Improve Teaching Quality

Fabricated design is systematic and sophisticated. Therefore, students can take advantage of teamwork when doing homework by putting forward their own fabricated structural design towards interior space through mutual discussion. Substitute teachers and designers with experience in fabricated design may provide further guidance to their structures and solve the relationship between appearance design and fabricated structure in a pointed way to further improve the plan. Good design may be used for development and production in manufacturers. The setup of interior design courses may enhance interaction between students and excavate their potential for students to carry out scientific and technological innovation in the research front related to the courses. As for different teaching contents and learning stages, teaching methods including discussion, demonstration and enlightenment will be adopted to improve students' learning initiative and guide them to independently and actively participate in the learning process as well as exploration and innovation, making them real subjects of study. Students are encourage to think actively and practice boldly to improve the teaching effect.

IV. Conclusion

The application of fabricated technology nowadays can not only improve the work efficiency, technological quality and environmental friendliness of interior decoration, but also manifest modernized professional design. Therefore, the combination between specialized courses on fabricated decoration and interior design can improve the course teaching effect and play a key role in improving students' course design ability. Besides, students are also allowed to flexibly apply their theoretical knowledge on textbooks to practical design projects, making them more innovative in future design and helping improve work efficiency. High-quality college students in design will meanwhile promote positive development of green architecture and industrialized fabricated interior construction in China.

活动总结

刘恒　王传顺

　　欣喜地看到了一场很有特色的学生作品展示，大家有不同的方向、不同的风貌，却都有深度的思考。这次"绿色营造"的主题特别有时代意义，也扩展了大家的思路。绿色是价值观的引入，是系统性解决问题的方法论，营造更是拓展了设计范畴，从设计到建造有了更多的可能性。整个毕业设计的过程中同学们和老师一起从选题研究，再到设计成果与表达一气呵成，思考也是越来越深入。

　　很多作品考虑到以人的使用为切入点，功能的复合化，甚至考虑到了运营过程；有的考量工艺和建造中的可能性，反推出设计的形式空间与语言；也有的以未来的可变性和弹性设计来展开，将功能、空间、装饰、家具一体化思考；还有的从环境中找到出发点，如场所、日照、通风、使用等，这些都特别有意义和价值。学生阶段的设计一定要强调概念的完整性和系统性，这里的确看到了不少好的设计。当然有的作品过于聚焦到一些形式本身，反而缺少设计解决问题的初心；也有的对装配概念过于局限到了一些节点上，显得过度工程化，有点放不开了，但对学生来说都是学习和历练。在汇报表达的把控和方式上，学生们也是对自己的又一次锻炼，这方面明显拉开了差距，如何在有限时间表达出方案的精髓是大家走向职业化的重要一步。整个毕业设计的过程也看出了指导老师们的悉心指导，让不少学生的作品有了质的提升，这也是师生共同互助的结果。

　　现代设计很多的机会都来自于选址和前期的策划，这次也很欣喜地看到了从这里产生的机会，看到在老城区里、在旧建筑中、在乡村、在未来互联网大潮下，等等的机会，让学生能在走向社会后具有更加敏锐的观察力和解决问题的能力，这些都是难能可贵的。

　　多校联合的方式，让学生和教师间能充分的交流、展现，校内校外导师交叉评论的方式给了创作者更多的思路和不同的视角。学会的良好组织也给毕业设计带来了新的活力。总之，这对所有参与者都是一次精彩的经历，作品有精彩也有不足，但对于学生们的成长意义非凡，在比对中学习、在实践中发现，这些都是职业生涯开始前的重要一笔。愿精彩明年继续！

Activity Summary

Liu Heng　Wang Chuanshun

　　I am glad to view the characteristic student works representation, which shows your different orientations and styles, and in-depth thinking. The topic of "Green construction" is of great significance of times, and broadens your thought. Green introduces the values, and is the methodology for systematic problem solution. Construction further expands the design scope, and brings more possibilities for design and construction. The students and teachers in the process of graduation design selected the subject, researched, and accomplished the design result and expression at a stretch, and thought in more and more comprehensively.

　　Many works took the approach of the use by people, and considered function compounding and even operation process; some considered the probability of process and construction and deducted the design pattern space and language in return; some started from the changeability and flexibility design in the future and thought about the integration of functions, space, decorations and furniture; some took the approach from the environment, such as place, daylighting, ventilation and use, which were all of special significance and value. The design for student stage must emphasize the completeness and systematicness of the concept. We did find plenty of good designs this time. Of course, some works excessively emphasized the form and lost the original intention to solve the issue in design; some restricted the assembly concept to some nodes, and made the design too engineering and a little bit restrictive. But it was an opportunity for the students to learn and practice. In regard to the control and method of report expression, the students also took the chance to practice. Thus the gap among the students became apparent. How to express the essence of the scheme in a limited time is an important step for the students to become professional. In the process of graduation design, the tutors also carefully guided the students, and helped many students to significantly improve the works. This was the result of cooperation of the teachers and students.

　　Many opportunities of modern design come from site selection and preliminary plan. This time we are glad to see the valuable opportunities here, including these in the city center, old buildings, rural area and future Internet wave, to let the students to hold more sharp outsight and ability to solve issues after entering the society.

　　The model of multiple college cooperation enables the students and teachers to thoroughly exchange and demonstrate. The model of cross assessment of the college internal and external tutors brings more thinking and different perspectives to the creators. The good organization by the institute also brings new vitality to graduation design. In conclusion, it is a wonderful experience for all the participants. Though the works contain virtues and shortcomings, they are of great significance in the growth of the students. They learn in comparison and have their findings in the practice, which is an important activity before their career starts. I look forward to a continuously wonderful activity next year!

刘恒

王传顺

教育研究

热点命题 纷呈特色

联合指导 服务需求

风采定格

『室内设计6+』2019（第七届）联合毕业设计

"Interior Design 6+"2019(Seventh Year) Joint
Graduation Project Event

一串足迹

学会主办 业界支持 高校协同 产学支撑
在设计教育创新的道路上留下探索的脚印

中国建筑学会室内设
2019（第七届）联合毕业设计

我 们 都 是 追 梦 人

计分会 "室内设计 6+"
开题报告汇报会（北京建筑大学）

「室内设计 6+」2019（第七届）联合毕业设计
"Interior Design 6+"2019(Seventh Year) Joint Graduation Project Event

风采定格

计分会"室内设计 6+"
中期检查汇报会（浙江工业大学）

风采定格

风采定格

中国建筑学会室内设

2019（第七届）联合毕业设计

「室内设计 6+」2019（第七届）联合毕业设计
"Interior Design 6+"2019(Seventh Year) Joint Graduation Project Event

232

计分会 "室内设计 6+"
答辩汇报会（南京艺术学院）

IID-ASC
中国建筑学会室内设计分会

中国建筑学会室内设计分会前身是中国室内建筑师学会，成立于 1989 年，是在中国建筑学会直接领导下、民政部注册登记的社团组织，是获得国际室内设计组织认可的中国室内设计师的学术团体。

分会的宗旨是引领室内设计行业的学术发展，联合全国室内设计师和相关资源，推动中国室内设计行业走向国际化舞台。

分会每年在全国各地举办学术活动，为设计师提供交流和学习的场所，同时也为设计师提供丰富的设计信息，加强室内设计行业国际间学术交流活动，促进中国室内设计行业更好更快地发展。

Institute of Interior Design-ASC

Six colleges display differently and this reflects a certain difference between them.The predecessor of the Institute of Interior Design-ASC (IID-ASC) is China Institute of Interior Architects. Since its establishment in 1989, IID-ASC has been the only authorized academic institution in the field of interior design in China.

IID-ASC aims to unite interior architects of the whole country, raise the theoretical and practical level of China's interior design industry, pioneer the Chinese characteristics of interior design, help interior architects play their social role, preserve the rights and interests of interior architects and foster professional exchanges and cooperation with international peers.

IID-ASC hold abundant and colorful academic exchanges every year, building a platform for designers to communicate and to study meanwhile update designer information of design industry to enhance the better and rapid development of interior design industry of China.

IID-ASC Secretarial is located in Beijing, taking charge of institute work.

同济大学

同济大学创建于 1907 年，教育部直属重点大学。同济大学 1952 年在国家院系调整过程中成立建筑系，1986 年发展为建筑与城市规划学院，下设建筑系、城市规划系和景观学系，专业设置涵盖城市规划、建筑设计、景观设计、历史建筑保护、室内设计等广泛领域。同济大学建筑与城市规划学院是中国大陆同类院校中专业设置齐全、本科生招生规模最大，世界上同类院校中研究生培养规模第一，具有全球性影响力的建筑规划设计教学和科研机构，是重要的国际学术中心之一。

同济大学室内设计教育起源于建筑系，同济大学建筑系于 20 世纪 50 年代就开始注重建筑内部空间的研究，1959 年曾尝试在建筑学专业中申请设立"室内装饰与家具专门化"。1986 年经国家建设部和教育部批准，同济大学建筑系成立了室内设计专业，1987 年正式招生，成为中国大陆最早在工科类（综合类）高等院校中设立室内设计专业的大学。1996 年原上海建材学院室内设计与装饰专业并入同济大学建筑系；2000 年原上海铁道大学装饰艺术专业并入同济大学建筑系。2009 年同济大学开始恢复建筑学专业（室内设计方向）的招生工作。2011 年建筑学一级学科目录下，设立"室内设计"二级学科。

同济大学建筑城市规划学院的教学理念为以现代建筑的理性精神为灵魂，以自主创造、博采众长的学术品格为本色，以当代技术与地域文化的并重交融为导向，以国际学科前沿的跟踪交流为背景。室内设计教学突出建筑类院校室内设计教学特色，强调理性精神，提出"以人为本、关注生态、注重环境整体观、时代性和地域性并重、融科学性和艺术性于一体"的室内设计观。

Tongji University

Tongji University, established in 1907, is a top university of China Ministry of Education. During the time of restructuring of the university and college systems in 1952,the Department of Architecture was formed at Tongji University , and in 1986 was renamed as the College of Architecture Urban Planning (CAUP). Currently CAUP has three departments: the Department of Architecture, the Department of Urban Planning, the Department of Landscape Design. The undergraduate program covers:Architecture, Urban Planning, Landscape Design, Historic Building Protection and Interior Design. CAUP is one of China's most influential educational institutions with the most extensive programs among its peers, and the largest body of postgraduate students in the world. Today, CAUP has been recognized as an international academic center with a global influence in the academic fields.

Tongji University's interior design education originated from the Department of Architecture which started to conduct interior space research in the 1950's.In 1959, it applied for the establishment of the "Interior Decoration and Furniture Specialty" within Architecture Discipline. In 1986,approved by the Ministry of Education and the Ministry of Construction, the "Interior Design Discipline" was formally founded. Starting to admit undergraduate students in 1987,Tongji University was one of two earlies thigh education institutions in mainland China to train interior design professionals in a University of science and technology.In 1996, the former interior design and decoration major of Shanghai Building Materials College was incorporated into the Department of architecture of Tongji University; in 2000, the former decoration art major of Shanghai Railway University was incorporated into the Department of architecture of Tongji University. In 2009, Tongji University began to resume the enrollment of architecture major (interior design direction). In 2011, the second level discipline of "interior design" was established under the first level discipline catalog of architecture.

"Interior Design" officially became the secondary discipline of the Architecture Discipline. In the same year, the "Interior Design Research Team" was established,providing even broader room for subject development. Tongji University's interior design education crystallizes its own characteristics,emphasizing rational thinking and proposing the interior design concept of"human centric, ecological consciousness, overall environmental perspective,equal time and regional characteristic significance,technology and art integration".

华南理工大学

　　华南理工大学位于广东省广州市，创建于1934年，是历史悠久、享有盛誉的中国著名高等学府。是中华人民共和国教育部直属的全国重点大学、首批国家"211工程""985工程"重点建设院校之一。

　　华南理工大学设计学院组建成立于2010年6月，现有工业设计、环境设计、信息与交互设计、服装与服饰设计等4个系。设计学院紧密依托华南理工大学雄厚的理工优势和深厚的人文底蕴，积极探寻与产业高度结合和国际化合作的道路，旨在打造享誉国内外设计创新人才培养和设计实践与服务的研究高地。

　　当前，设计学院紧紧把握设计创意产业的发展契机，不断创新 教育理念，大胆探索设计创新人才培养模式，以"技术创新引领、文化创意引领、产业转型引领、可持续发展引领"为建设目标，拥有"创意与可持续设计研究院"以及"当代艺术空间""设计实验与实践公共平台""跨学科拔尖创新人才培养试验区"和"腾龙研发中心""文化艺术与创意产业研究中心""中国民间艺术研究中心""陶瓷文化研究所"等一系列产学研平台，力争建设成为国内领先、有国际影响力的设计学院，从而支撑、引领国家和广东设计产业发展。

South China University of Technology

　　South China University of Technology(SCUT), located in Guangzhou City, Guangdong Province, was founded in 1934. It is a well-known Chinese university which has a long history and enjoys a high reputation. It is a national key university directly under the Ministry of Education of the People's Republic of China, one of the first national "211 project" and "985 project" key construction of colleges and universities.

　　The design institute of SCUT was founded in June 2010, with majors including industrial design, environmental design, information and interaction design, clothing and apparel design. The design institute closely relies on the strong advantage of technology and deep cultural heritage of SCUT and actively explores the way of highly industry integration and international cooperation, aiming to create famous heights for domestic and foreign design innovative talent training as well as design practice and service.

　　At present, the design institute grasps the development opportunity in design creativity industry, constantly renews education idea and makes bold exploration in design innovation personnel training mode. With "leading technology innovation, leading culture innovation, leading industry transformation and sustainable development" as the construction goal, a series of production and research platform including "creative and sustainable design and research institute" "space of contemporary art" "public platform of design experiment and practice" "interdisciplinary top creative talents cultivation test area" and "Teng Long research&development center" "cultural art and creative industry research center" "Chinese folk art research center" "ceramic culture research institute" has been built, striving to become the domestic leading design institute with international influence, so as to support and lead the design industry development in Guangdong and across the country.

哈尔滨工业大学

哈尔滨工业大学隶属于国家工业和信息化部，是首批进入国家"211工程""985工程"和首批启动协同创新"2011计划"建设的国家重点大学。1920年，中东铁路管理局为培养工程技术人员创办了哈尔滨中俄工业学校——即哈尔滨工业大学的前身，学校成为中国近代培养工业技术人才的摇篮。学校已经发展成为一所特色鲜明、实力雄厚，居于国内一流水平，在国际上有较大影响的多学科、开放式、研究型的国家重点大学。

哈尔滨工业大学建筑学学科是我国最早建立的建筑学科之一，历经90余载风雨砥砺。建筑学院建筑学科现有建筑学、城乡规划、风景园林、环境设计4个本科专业和建筑学、城乡规划学、风景园林学3个一级学科点和设计艺术学二级学科硕士点。已获得建筑学、城乡规划学和风景园林一级学科博士、硕士授予权，以及设计学二级学科硕士授予权，还设有建筑学一级学科博士后科研流动站。建筑学院始终秉持严谨治学、精于耕耘的文化精神，打造了一支朴实敬业、有特色、有能力、肯奉献的优秀教师团队。在本科教学、研究生培养及科学研究方面，特色鲜明，成绩显著。在寒地公共建筑设计、地域建筑设计、寒地建筑技术、建筑历史与理论、寒地城市规划与城市设计、寒地环境艺术设计等诸多方向上，均形成自己的学术特色。

Harbin Institute of Technology

Harbin Institute of Technology affiliates to the Ministry of Industry and Information Technology, and is among the first group of the national key universities to enter the national"211Project""985 Project" and to start the collaborative innovation "2011 plan". In order to train engineers, the Mid east railway authority founded the Harbin Sino Russian school in 1920, the predecessor of Harbin Institute of technology, which becomes the cradle of China's modern industry and technical personnel.The School has evolved into a distinctive, powerful, first class national key university, which is multidisciplinary, open, researchful and with international influence.

The discipline of Architecture in Harbin Institute of technology is one of the earliest architectural subjects in China, with more than 90 years'ups and downs. The school of Architecture has 4 undergraduate disciplines, including Architecture, Urban and Rural Planning, Landscape Architecture,Environmental Design, and 3 first-level disciplines, including Architecture, Urban and Rural Planning,Landscape Architecture, and secondary master's disciplines in Design and Arts. We have the first-level doctorate and master's authorization in Architecture, Urban and Rural Planning and Landscape Architecture, and secondary-discipline master's authorization in Design, and Post-doctoral Research Institute on architectural first-level discipline.With the cultural spirits of rigor and diligence,The school of Architecture has created a devoted, distinctive, qualified and dedicated teachers' team.We have gained distinctive and outstanding achievements in undergraduate teaching, postgraduate education and scientific research, and have formed our own academic characteristics in the Design of Public Buildings in Cold Region, Regional Architecture, Building Technology in Cold Region, Architectural History and Theory, Urban Planning and Designing in Cold Region and Environmental Design in Cold Region.

西安建筑科技大学

西安建筑科技大学坐落在历史文化名城西安，学校总占地4300余亩，校园环境优美，办学氛围浓郁。学校办学历史源远流长，其办学历史最早可追溯到始建于1895年的北洋大学，积淀了我国近代高等教育史上最早的一批土木、建筑、环境类学科精华。1956年，时名西安建筑工程学院。1959年和1963年，曾先后易名为西安冶金学院、西安冶金建筑学院。1994年3月8日，经国家教委批准，更名为西安建筑科技大学，是公认的中国最具影响力的土木建筑类院校之一及原冶金部重点大学。

西安建筑科技大学是以土木、建筑、环境、材料学科为特色，工程学科为主体，兼有文、理、经、管、艺、法等学科的多科性大学。学校现有16个院（系），其60个本科专业面向全国第一批招生，有权招收保送生，实行本硕连读。艺术设计本科专业为陕西省特色专业。

西安建筑科技大学艺术学院成立于2002年4月，是由建筑学院的艺术设计专业和摄影专业本科生、机电工程学院工业设计专业本科生和新成立的雕塑专业及各专业教师组建而成。学院现有艺术设计、工业设计、摄影、雕塑、会展艺术与技术5个本科专业，在校本科生1200余人。艺术设计专业被评为"国家级特色专业""省级名牌专业"。 学院集聚了包括建筑、规划、景观等在内的多学科的研究人才，学科团队长期致力于西部地区地域文化研究，承担了多项国家、省部级基金课题。艺术学院积极主办（承办）国家级学术、学科建设会议；邀请国际、国内知名教授来我院进行学术交流；制定管理办法，并设立专项基金，鼓励青年教师和优秀博士生开展学术交流、国际（内）合作研究，与欧洲、亚洲地区的多所大学建立了友好合作关系。

学院以学生全面发展为培养目标，注重学生综合素质的提高，依托各类学生组织载体和平台，开展形式多样的课外活动。注重加强学术交流与互动，邀请学者、专家和社会知名人士来我院举办讲座和专题报告，开阔学生视野，改善学生知识结构，培养学生的科技、人文精神。组织学生积极参与学科

Xi'an University of Architecture and Technology

Located in the historical and cultural city Xi'an, covering an area of 4300 acres, Xi'an University of Architecture and Technology has beautiful campus environment and academic atmosphere. This university has quite a long history, which can be dated back to the Northern University, founded in 1895. Since then, in the higher education history of modern China, this university has been accumulating the first batch of disciplines essence in civil engineering, construction and environmental class. In 1956, this university was named as Xi'an Institute of Architectural Engineering. In 1959 and 1963, it was renamed as Xi'an Institute of Metallurgy and Xi'an Institute of Metallurgy and Construction. On March 8, 1994,approved by the State Board of Education, it was renamed as Xi'an University of Architecture and Technology and was recognized as one of China's most influential civil engineering colleges and the key university of the former Ministry of colleges and the key university of the former Ministry of Metallurgical.

Featured by civil engineering, construction,environment and materials science, engineering disciplines as the main body, Xi'an University of Architecture and Technology is a multidisciplinary university also with liberal arts, science, economics, management, arts, law and other disciplines. The university has 16 departments, 60 undergraduate programs so it can launch the first batch of undergraduate enrollment. It also has the right to recruit students by recommendation and the right of implementation of Accelerated Degree. Undergraduate art and design program is the featured major in Shaanxi Province.

Founded in April, 2002, Xi'an University of Architecture and Technology was established by the undergraduates from the major of art design and photography and from mechanical and electrical engineering industrial design and the relevant teachers from newly established sculpture and other specialties. The current undergraduate majors in this college include art and design, industrial design, photography, sculpture, exhibition art and technology, with more than 1,200 undergraduate students. Art Design was named "national characteristic specialty""provincial famous professional". This university has gathered many multidisciplinary researchers,including architecture, planning, landscape, etc. All these research teams have a long history of working towards the research of western region cultures, through undertaking many national and provincial funds subjects. The Arts College has actively organized (or as the contractor) the national academic, discipline-building meetings; inviting international and domestic famous professors to come for academic exchanges. It also has developed management approach, and set up a special fund to encourage young teachers and outstanding doctoral students to carry out academic exchanges and international (inside) collaborative researches. In the meantime it has established friendly and cooperative relations with the universities in Europe, Asia and

竞赛，指导、鼓励学生从事科研活动，在国内刊物上发表各类论文。学院调动教研室、资料室、实验室进行多方互动，通力合作，构建了教学、科研、学生三位于一体的开放性实验（工作）平台。学院培养的学生深受用人单位欢迎，毕业生供不应求。

other countries.

The university has taken the overall development of students as its training objectives, the improvement of the overall quality of them as the aim to focus on. Relying on various student organizations carrier and platforms,the university has carried out various forms of extracurricular activities. And also it has focused on strengthening academic exchanges and interaction, inviting scholars, experts and celebrities to come to listen to the lectures and presentations,which can broaden the students' horizons, improve their knowledge structure and culture their spirits of science, technology and humanities. In other ways, the university organized the students to actively participate in academic competitions, and guided or encouraged students to engage in research activities, and many students have published various papers in the national magazines. The college has transferred departments, libraries, laboratories and paid multi-interactive efforts or work together to build a teaching-research-student trinity open experiment (work) platform. The graduates trained by the college have been welcomed by employers and the graduates are in short supply.

北京建筑大学

北京建筑大学是北京市和住房城乡建设部共建高校、教育部"卓越工程师教育培养计划"试点高校和北京市党的建设和思想政治工作先进高校，是一所具有鲜明建筑特色、以工为主的多科性大学，是"北京城市规划、建设、管理的人才培养基地和科技服务基地""北京应对气候变化研究和人才培养基地"和"国家建筑遗产保护研究和人才培养基地"，是北京地区唯一一所建筑类高等学校。

学校源于 1907 年清政府成立的京师初等工业学堂。1977 年学校恢复本科招生，1982 年被确定为国家首批学士学位授予高校，1986 年获准为硕士学位授予单位。2011 年被确定为教育部"卓越工程师教育培养计划"试点高校。2012 年"建筑遗产保护理论与技术"获批为服务国家特殊需求博士人才培养项目，成为博士人才培养单位。2014 年获批设立"建筑学"博士后科研流动站。2015 年 10 月北京市人民政府和住房城乡建设部签署共建协议，学校正式进入省部共建高校行列。2016 年 5 月，学校"未来城市设计高精尖创新中心"获批"北京高等学校高精尖创新中心"。2017 年获批推荐优秀应届本科毕业生免试攻读研究生资格。2018 年 5 月，获批博士学位授予单位，建筑学、土木工程获批博士学位授权一级学科点。

学校有西城和大兴两个校区。目前，学校正在按照"大兴校区建成高质量本科人才培养基地，西城校区建成高水平人才培养和科技创新成果转化协同创新基地"的"两高"发展布局目标加快推进两校区建设。与住建部共建中国建筑图书馆，是全国建筑类图书种类最为齐全的高校。

学校坚持立德树人，培育精英良才。现有各类在校生 11842 人，已形成从本科生、硕士生到博士生和博士后，从全日制到成人教育、留学生教育全方位、多层次的办学格局和教育体系。多年来，学校为国家培养了6 万多名优秀毕业生，他们参与了北京 60 年来重大城市建设工程，成为国家和首都城市建设系统的骨干力量。学校毕业生全员就业率多年来一直保持在 95% 以上，2014 年进入"全国高校就业 50 强"行列。

学校面向国际，办学形式多样。学校始

Beijing University of Civil Engineering and Architecture

Beijing University of Civil Engineering and Architecture is a university co-constructed by Beijing City and the Ministry of Housing and Urban-Rural Development, a pilot university of the "Excellent Engineer Training Program" initiated by the Ministry of Education, a university active in the Party building and ideological and political work of Beijing City, an engineering-based multiversity with outstanding architectural features, a "base for training of planning, construction and management personnel and a high-tech service base" in Beijing, a "Climate change treatment research institute and personnel training base" in Beijing, a "national architectural heritage conservation and research and personnel training base", and the only university of architecture in Beijing.

The university was formerly known as Beijing Primary Technical School, which was founded in 1907. The university resumed undergraduate admissions in 1977, was identified as one of the first undergraduate universities in 1982, authorized to award master's degrees in 1986, and identified as a pilot university of the "Excellent Engineer Training Program" initiated by the Ministry of Education in 2011. In 2012, its "architectural heritage conservation theory and technology" was approved as a national special doctoral talent training program, making it a doctoral talent training institution. In 2014, the university built a center for post-doctoral studies on "architecture". In October 2015, the Beijing Municipal Government and Ministry of Housing and Urban-Rural Development signed a co-construction agreement, listing the university among the universities co-constructed by the province and ministry. In May 2016, the university's "Sophisticated Innovation Center for Future City Design" was renamed "Beijing Sophisticated Innovation Center for Institutions of Higher Education". In 2017, the university became eligible to recommend fresh undergraduate graduates to receive postgraduate studies without sitting for the entrance examinations. In May 2018, the university was approved as a doctor's degree granter, and its architecture and civil engineering were upgraded to first-level disciplines for doctor's degree.

The university has two campuses, which are located in Xicheng District and Daxing District respectively. Currently, the university is accelerating the construction of the two campuses to build the "campus in Daxing District into a base for high-quality undergraduate talent training and the campus in Xicheng District into a base for high-level talent training and transformation of technological innovation achievements as well as collaborative innovations". The Chinese Architectural Library in it, built with the Ministry of Housing and Urban-Rural Development, is a library with the fullest range of architectural books in China.

The university insists on educating students by virtue and raising elites. Now the university has 11842 students, including undergraduate, postgraduate, doctoral and post-doctoral students, and a multi-level educational system covering full-time teaching, adult education and international students education. Over the years, the university has

终坚持开放办学战略，广泛开展国际教育交流与合作。目前已与美国、法国、英国、德国等 28 个国家和地区的 63 所大学建立了校际交流与合作关系。

站在新的历史起点上，学校将以党的十九大精神为指引，深入学习贯彻习近平新时代中国特色社会主义思想，按照"提质、转型、升级"的工作方针，围绕立德树人的根本任务，全面推进内涵建设，全面深化综合改革，全面实施依法治校，全面加强党的建设，持续增强学校的办学实力、核心竞争力和社会影响力，以首善标准推动学校各项事业上层次、上水平，向着把学校建设成为国内一流、国际知名、具有鲜明建筑特色的高水平、开放式、创新型大学的宏伟目标奋进。

provided the society with more than 60000 excellent graduates, who have participated in the major urban construction projects in Beijing in the past 60 years, becoming the backbone of the national and Beijing's urban construction systems. Its graduate employment rate has been above 95% for many years, and ranked among "China's top 50" in 2014.

The university offers forms education in the face of the world. It always insists on open schooling, and conducts international educational exchanges and cooperation extensively. Currently, the university has built an intercollegiate exchange and cooperation relationship with 63 universities in 28 countries and regions such as America, France, Britain and Germany.

Standing on a new historical starting point, the university will profoundly learn and implement Xi Jinping's thought on socialism with Chinese characteristics for a new era under the guidance of the spirit of the 19th CPC National Congress, comprehensively promote the connotation construction, fully deepen the comprehensive reform, fully implement schooling by law, fully enhance the Party building, constantly strengthen its schooling strength, core competitiveness and social influence around the fundamental task of educating students by virtue under the direction of the work policy of "quality improvement, transformation and upgrading", and increase the level of its various undertakings by considering itself the greatest philanthropist, in a bid to make itself a domestic first-class and internationally known high-level, open innovative university with outstanding architectural features.

风采定格

南京艺术学院

南京艺术学院是我国独立建制创办最早并延续至今的高等艺术学府。下设 14 个二级学院，27 个本科专业及 50 个专业方向。拥有艺术学学科门类下设的艺术学理论、音乐与舞蹈学、戏剧与影视学、美术学以及设计学全部 5 个一级学科的博士、硕士学位授予权及博士后科研流动站。

南京艺术学院从 2005 年开设了展示设计本科专业和硕士专业研究方向；2008 年该专业并入工业设计学院，2011 年会展艺术与技术专业作为独立的二级学科获得国家教育部的正式批准，2012 年该专业又被归为设计学类，成为"艺术与科技"专业。南京艺术学院工业设计学院的艺术与科技（展示设计）专业以学生为中心，以学术为导向，以实践为手段，以发展为目标，通过近 10 年的发展，已经逐步形成知识融贯、结构合理、连贯而开放的模块化专业课程体系和走向现代化、全球化的课程内容。旨在为文化部门、博物馆部门、大中型展馆、设计团体、旅游部门、会展机构等单位培养具有一定的理论素养，专业知识合理，专业特点突出，具备问题导入、市场导入和文化导入的整合设计和研究能力，以及高度艺术造型及表达能力的专业设计人才。

Nanjing University of the Arts

Nanjing University of the Arts is one of the earliest arts institutions in China. It consists of 14 schools, 27 undergraduate majors and 50 major directions. It has Master's and Doctoral degrees and Post-doctoral stations in 5 disciplines under the first-class discipline of the arts: Arts Theory, Music and Dance, Drama and Film, Fine Arts Theory and Design Theory.

The professional background: in 2005, Display Design was set up as a major direction in undergraduate level and a research direction in master program in Nanjing University of the Arts; in 2008, it was incorporated into industrial design major as its one direction in School of Industrial Design; in 2011, it was approved by the Ministry of Education as an independent sub-discipline in then national disciplinary classification in undergraduate education; in 2012, it was classified into the first-class discipline of design with a new major name of "Art and Technology ".Through nearly 10 years of efforts by adhering to the principle that is students centered, academy-oriented, practice focused and development-guided, Art and Technology (Display Design) major has formed a coherent and open modular curriculum system of coherent knowledge and rational structure supported by modernization and globalization oriented course contents. The major is to cultivate professional design talents for the cultural sector,the museum sector, medium and large exhibition halls, the design community, the tourism sector, exhibition and other institutions. The graduates of this major are to have the ability to do design and research in the manner of integrating question, market and culture. And they are also to be cultivated as talents with the capacity of high-level artistic formation and excellent expression, as well as rational expertise structure and outstanding professional features.

浙江工业大学

浙江工业大学是一所教育部和浙江省共建的省属重点大学，其前身可以追溯到1910年创立的浙江中等工业学堂。经过几代人的艰苦创业和不懈奋斗，学校目前已发展成为国内有一定影响力的综合性的教学研究型大学，综合实力稳居全国高校百强行列。

2013年浙江工业大学牵头建设的长三角绿色制药协同创新中心入选国家2011计划，成为全国首批14家拥有"2011协同创新中心"的高校之一。目前学校有本科专业68个，硕士学位授权二级学科101个，博士学位授权二级学科25个，博士学位授权一级学科5个，博士后流动站4个。学科涵盖哲学、经济学、法学、教育学、文学、理学、工学、农学、医学、管理学、艺术学等11个大门类。学校师资力量雄厚，拥有中国工程院院士2人、共享中国科学院和中国工程院院士3人、国家级有突出贡献中青年专家6人、国家级教学名师3人、国家杰出青年基金获得者3人、中央千人计划入选者2人、教育部长江学者特聘教授1人、教育部创新团队1个、国家级教学团队2个、各类国家级人才培养计划入选者26人次。浙江工业大学坚持厚德健行的校训，把提高教育质量放在突出位置，努力培养能够引领、推动浙江乃至全国经济和社会发展的精英人才。

Zhejiang University of Technology

Zhejiang University of Technology is a key comprehensive college of the Zhejiang Province; its predecessor can be traced back to the founding in 1910 as Zhejiang secondary industrial school.After several generations' hard working and unremitting efforts, the school now has grown to be a comprehensive University in teaching and researching which is very influential. The comprehensive strength ranks the top colleges and universities. In 2009, Zhejiang province people's government and the Ministry of Education signed a joint agreement; Zhejiang University of Technology became the province ministry co construction universities.

In 2013 Zhejiang University of Technology led the construction of Yangtze River Delta green pharmaceutical Collaborative Innovation Center which was selected for the national 2011 program,to become one of the first 14 of 2011 collaborative innovation center. There are 68 undergraduate schools; 101 grade-2 subjects of master's degree authorization; 25 grade-2 subjects of doctor's degree authorization; 4 postdoctoral research stations. Subjects include philosophy, economics, law,education, literature, science, engineering, agriculture, medicine, management, arts and other 11 categories. School teacher is strong. There are 2 Chinese academicians of Academy of Engineering,sharing 3 academicians of Chinese Academy of Sciences and Academy of Engineering; 6 national young experts with outstanding contributions; 3 National Teaching Masters, 3 winners of national outstanding youth fund, 2 people were selected to central thousand person plan, the Ministry of education,1 professor of the Yangtze River scholars, 1 innovative team of Ministry of Education, 2 national teaching teams, and 26 person were selected to all kinds of national personnel training plans.Zhejiang University of Technology adheres to its motto "Profound accomplishment and invigorating practice. Accumulate virtues and good practice." To improve the quality of education in a prominent position, and strive to cultivate to lead, promote Zhejiang and even the country's economic and social development of elite talent.

致 谢

Acknowledgements

中国建设科技集团股份有限公司

中国建筑设计咨询有限公司 绿色建筑设计研究院

中国建筑标准设计研究院有限公司

中国建筑设计研究院有限公司

华东建筑集团股份有限公司 上海现代建筑装饰环境设计研究院有限公司

南京观筑历史建筑文化研究院

杭州国美建筑设计研究院有限公司

上海全筑建筑装饰集团股份有限公司

常州霍克展示系统股份有限公司

江苏锦上装饰设计工程有限公司

亚洲城市与建筑联盟

东易日盛家居装饰集团股份有限公司